SOLAR HEATING SYSTEMS

Analysis and Design with the Sun-Pulse Method

GORDON TULLY

An Energy Learning Systems Book

McGRAW-HILL BOOK COMPANY

New York St. Louis San Francisco Auckland Bogotá Hamburg Johannesburg London Madrid Mexico Montreal New Delhi Panama Paris São Paulo Singapore Sydney Tokyo Toronto

Library of Congress Cataloging in Publication Data

Tully, Gordon.

Solar heating systems.

(An Energy learning systems book)

Bibliography: p.

Includes index.

1. Solar heating. 2. Solar collectors. I. Title.
II. Series: Energy learning systems book.

TH7413.T84 697'.78 80-20238

ISBN 0-07-065441-7

1234567890 KPKP 8987654321

CONTENTS

PREFACE vii

INTRODUCTION ix

Chapter 1 A SIMPLE SOLAR HEATING SYSTEM 1

The Basic Diagram 1
System Temperatures 3
How the System Copes with Sun-Pulses 4
A Preview of System Operation 7

Chapter 2 COLLECTOR LOCATION 9

Sun Angles 9
Shading and Site Planning Issues 14
Guidelines for Designers 20
Retrofit Installation Checklist for Shading and Collector Location 20
New Installation Checklist for Shading and Collector Location 22

Chapter 3 THE SUN—PULSE 25

The Clear Day True-South Sun-Pulse 25
Deriving the Basic Sun-Pulse Data 27
Cloudy Day Sun-Pulses 31

The Correction for Off-Angle Reflection 35
Non-South Sun-Pulses 37
Shaded Sun-Pulses 37
Tilt and Ground Reflectance 39
A Summary of the Sun-Pulse Concept 41

Chapter 4 THE COLLECTOR 44

The Output Equation 44
Cover and Absorption Losses 45
Collector Heat Loss 47
Collector Heat Transfer Coefficient 54
Graphing Collector Performance 62
Relating Collector Performance to the Sun-Pulse 70
A Caution about Averaging 72
Controlling Active Collectors 73
Guidelines for Designers 78
Collector Checklist for Designers 79

Chapter 5 HEAT EXCHANGE 83

Heat Exchangers 83
Designing with Heat Exchangers 85
Guidelines for Designers 94
Exponential Decay and System Time Constants 96
The Relationship between Capacitance and Conductance 99

Chapter 6 HEAT LOSS AND INSULATION 104

The Building as a Heat Exchanger 104
A Closer Look at Heat Loss and the Effects of Occupancy 109
Guidelines for Designers 117

Chapter 7 SUPPLY AND DEMAND: SYSTEM BEHAVIOR AS A FUNCTION OF TIME 121

Time Variation in Supply and Demand 121
Monthly Supply and Demand 125
Short-Term Supply and Demand During Nearly 100% Solar Periods 132
Phase-Lag (Seasonal) Storage 134

System Controls I: Low Limits on Delivery Temperature 137
System Controls II: Night-Time Thermostat Set-Back 139
System Controls III: Controlling the Back-up System 142
Stratification in Water Storage 144
Rock Stores 146
Living in a Collector 149
A Brief Note on heat Pumps 158

Chapter 8 WATER HEATING 161

Supply, Demand, and Storage 161
Energy Conservation and Water Heating 163
Availability of Data 167
Guidelines for Designers 170

Chapter 9 SYSTEM OPTIMIZING 172

Defining the Optimum 172
Payback 174
Establishing System Cost 175
Determining System output 176
Guidelines for Designers: General Principles of System Optimizing 177
Guidelines for Designers: An Example of Optimizing 182

Chapter 10 DESIGNING A SIMPLE ACTIVE SOLAR HEATING SYSTEM 186

The Design of Drain-Back Systems 186
Drain-Back System Details 194
Pipe Sizing and Pump Selection Procedure for Drain-back Systems 200
Guidelines for Designers: A System Design Checklist 204

BIBLIOGRAPHY 206

GLOSSARY 210

INDEX 217

PREFACE

In the rapid development of solar heating, the handful of investigators working in 1973, when Massdesign began its involvement with solar heating, has turned into an army. Like all armies, this one has produced its share of paper, perhaps as a secondary hedge against an energy crisis—have we reached the magic day when the energy produced by solar heaters equals the energy available by burning the paper written about them? Most of the writing falls into three classes: theoretical and technical studies, usually dealing with a small special problem, often ignoring the rest of the industry, couched in the most technical possible language and spiced with accurate Greek symbolism; "soft" books of solar heating lore scanning the field and acting as primers to help the newly interested understand the basic ideas; and a growing list of "nuts and bolts" books, which take system design down to the level of actual installation and maintenance. Curiously lacking is a clear exposition of the principles which govern the behavior of these systems. This book attempts to fill this void in the literature.

Since a low-temperature tap-water drain-back system is used throughout as the basic system model, other system types have been dealt with as subtypes or variants to this basic design. In particular, air and passive systems are not treated in the same detail as liquid systems. Most of the system dynamics apply equally well to these important system types; the primary differences lie in storage. While water systems can successfully be considered to have homogeneous storage, both rock bins and intrinsic passive storage require that heat transfer within the store be considered. In this book, such complexities could be discussed only in general terms. The author expects that a simple theory, amenable to compact computer programs such as the Sun-Pulse II™ Simulation developed by the author, will be derived in the next year or two.

Neither cooling (naturally, this being a book about heating) nor photovoltaics is considered in detail, although many of the same systems principles derived for solar heating apply equally well to these subjects.

The book is written for people who are not afraid of pursuing and understanding subtle distinctions in the design of systems, but who do not have ready access to the language of calculus. While it is cumbersome to explain such concepts as exponential decay and heat exchanger effectiveness without using differential equations, it is a worthwhile goal to make these ideas accessible to the less "numerate," and to those who once could integrate and differentiate, but now can only expostulate and consider.

The book is organized in five parts:

- *Chapter 1* introduces the material, and provides a *basic description* of the system to be examined.
- *Chapters 2, 3 and 4* deal with the *supply side* of the system: the mechanism for harvesting heat from the sunlight, and the nature of the sun-pulse.
- *Chapter 5* deals with the general principles of *heat exchange*, as they apply to the overall system.
- *Chapters 6 through 9* form the heart of the book, dealing with the *relationships between supply and demand.* They cover the nature of the demand, principles of matching system with load, cost optimizing, and the behavior of each part of the system over time.
- *Chapter 10* describes in detail the *drain-back system* used as the model of a simple system throughout the book, including a discussion of the fluid dynamics of such a system.

Engineers experienced in the theory of collectors and heat exchangers may still wish to look at our diagrams. The author has tried throughout the book to reexamine every idea and "truth" current in the industry. The result may be very helpful to teachers searching for graphs and metaphors to explain these sometimes confusing but basically simple issues to students. English units are the basic ones used, since the book is aimed primarily at the construction industry, with a table of SI conversions included at the end of the book.

INTRODUCTION

In its six years of experience with solar building and system design, Massdesign, with whom the author is associated, has tried both to innovate and to live within the complex restraints of the building industry. The search for simplicity, economy, and reliability has led to new ways of thinking about solar-heated buildings, and this book is the result of that search. Along the way, we were fortunate enough to have a continuing dialogue with Acorn Structures, Inc., of Acton, Massachusetts, an established firm manufacturing contemporary homes. The firm's founder, John Bemis, led us toward many of our insights and made available to us the detailed results of a three-year computer monitoring of the Acorn Solar Test House. The firm's system, one of the most elegant and inexpensive systems available, pioneered many daring simplifications in liquid-cooled systems, and it was on the Acorn model that many of our projects were designed.

The original reason for writing this book was our frustration with, of all things, the prevailing views toward thermodynamics within the solar industry. Except for Will Hapgood at Raytheon, who designed the Acorn system along with John Bemis, few if any investigators shared our views about the advantages of large stores, large-load heat exchangers, and systems with low operating temperatures. One forever heard the designer or owner brag about how high a temperature his system could deliver. Granted that a large branch of the industry is chasing the elusive low-cost, high-temperature system for use in industrial process heating and air conditioning, it remains clear that high or even moderately high temperatures in most space heating applications are not only unnecessary but positively inconvenient.

The principles developed to defend and explain our very-low-temperature systems turn out to work superbly well in explaining all sorts of systems. This is helpful, since more sophisticated devices such as vacuum tube collectors might after all be successful, economical, and useful in solving simple heating problems. At the very least, we can expect a full range of system operating temperatures to persist on the market, each filling an important role. While the author might advocate very-low-temperature systems where applicable, in many cases higher temperatures are required. The optimization principles in this book will allow a designer to cope with any situation and arrive at the best solution.

Many of the ideas in the book are derived from Massdesign's earlier book, *Solar Heated Houses for New England*, published by the firm and prepared under a grant by the AIA Research Corporation. The principal author of that book, Hugh Adams Russell, provided us with many basic concepts, presentation methods, and the general impetus to write a book in the first place. Our friend and first solar client, Mark Hyman, led us to low-temperature systems by his experiment with a 100-percent system for his former house in Waltham, Massachusetts, for which Massdesign was the architect.

The book's early manuscript was read by Bill Shurcliff, Shawn Buckley, Tim Johnson, Bob Smith, Michael McClintock, Jan Kreider, Dr. Eric Farber, and Jeffrey Cook. All of them were as encouraging as the dickens, given the state of the manuscript at that point; and all corrected many of the numerous faults they found. Those remaining are the author's responsibility.

John Buchanan, who worked out the preliminary versions of the author's Sun-Pulse II Simulation helped the author clarify many of the system dynamics concepts. His ideas about the sun-pulse were seminal.

The book would never have been finished without the constant encouragement of Bob Entwistle of Energy Learning Systems, co-publisher. Woodsie Entwistle's assistance with the artwork was invaluable.

Sanford Sillman, who wrote the initial version Sun-Pulse II Simulation in record time, is also responsible for clarifying and expanding the author's ideas about seasonal storage, based upon his thesis done at MIT.

Finally, Tudor (Jerry) Ingersoll and Ralph Bennett, the author's coprincipals at Massdesign, and coauthors of Massdesign's first book, exhibited unique fiscal tolerance, emotional support, and good ideas.

1

A SIMPLE SOLAR HEATING SYSTEM

THE BASIC DIAGRAM

In order to explain the detailed behavior of a solar heating system in simple terms, we shall describe the simplest possible system first and add the complexities later. The diagram below shows such a system: an open tapwater, drain-back type system installed in a residence. This system might include many additional elements and we will discuss some of these later. The heart of the system, however, is shown in Fig. 1.1. Its major components are:

- The collector along with its associated pump, which converts solar radiation (visible light, near UV and IF) into heat, and then transfers the heat into a stream of water
- The store, which receives the heated water from the collector, returns water to be heated, delivers heated water to the space-heating load heat exchanger, and receives cooler water from the heat exchanger
- The space-heating load heat exchanger along with its associated pump, which transfers heat from the store to the interior of the house
- The building itself, which transfers heat from the interior air of the house to the outdoors
- Other heat sources, which include not only the back-up heater, but solar gain, and the "wild," "free," or "miscellaneous" heat generated as a byproduct of occupying the building
- A coil, immersed in storage, which pre-heats domestic water
- A differential controller to operate the collector

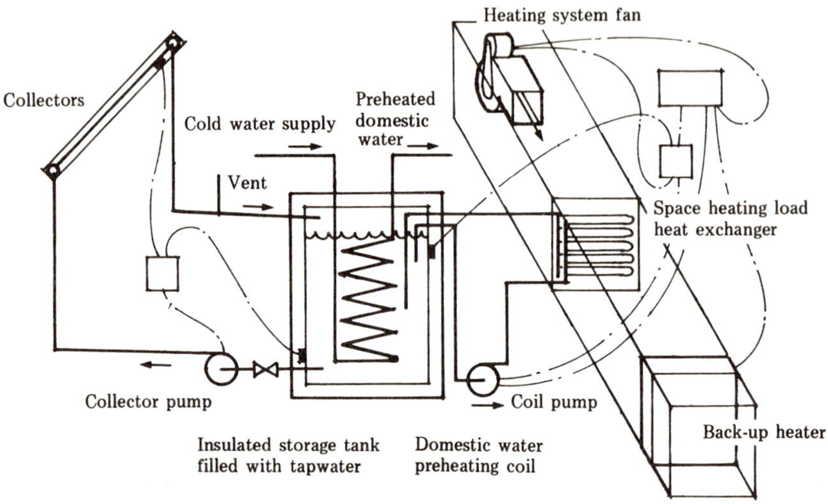

Fig. 1.1 Schematic diagram of a basic open-loop, tapwater, drain-back solar heating system with forced warm air distribution and domestic water preheat coil.

Fig. 1.2 Photos of the original test house and garage-mounted collectors at the Acorn Structures, Inc., plant in Acton, MA. The building has since been replaced by a hybrid model home.

- A thermostat to operate the load coil pump and fan, and the back-up heater
- A back-up heating device.

The system manufactured by Acorn Structures, of Acton, Massachusetts, is a good example of such a system. The Acorn Test House was used as a means of testing many of the concepts used in the book, since it was well instrumented, and many of these concepts were invented by the designers of the Acorn system.

SYSTEM TEMPERATURES

It is useful to modify the simple schematic diagram shown above, in order to place each element at its appropriate *temperature*. To understand how solar-heating systems work and thereby to make more intelligent choices of systems or components, it is necessary to understand how heat flows into and out of the system. This in turn requires understanding the temperatures of the various parts, and the ease with which heat flows among them. In Figs. 1.3, 1.4 and 1.5, the basic system diagram is redrawn to show system temperatures.

The arrows running down from the collector, the store, and the house represent the losses from the system. (In the Acorn house, the

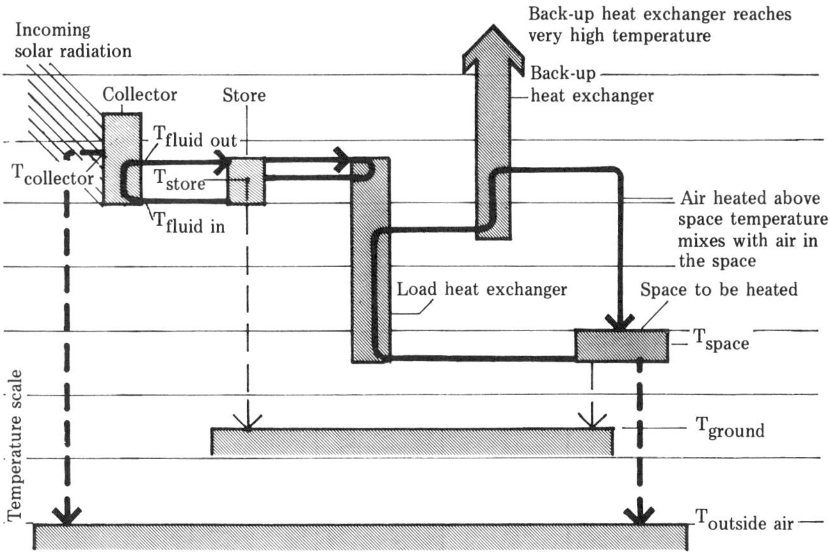

Fig. 1.3 Basic drain-back system schematic showing relative temperatures of the components (temperature schematic).

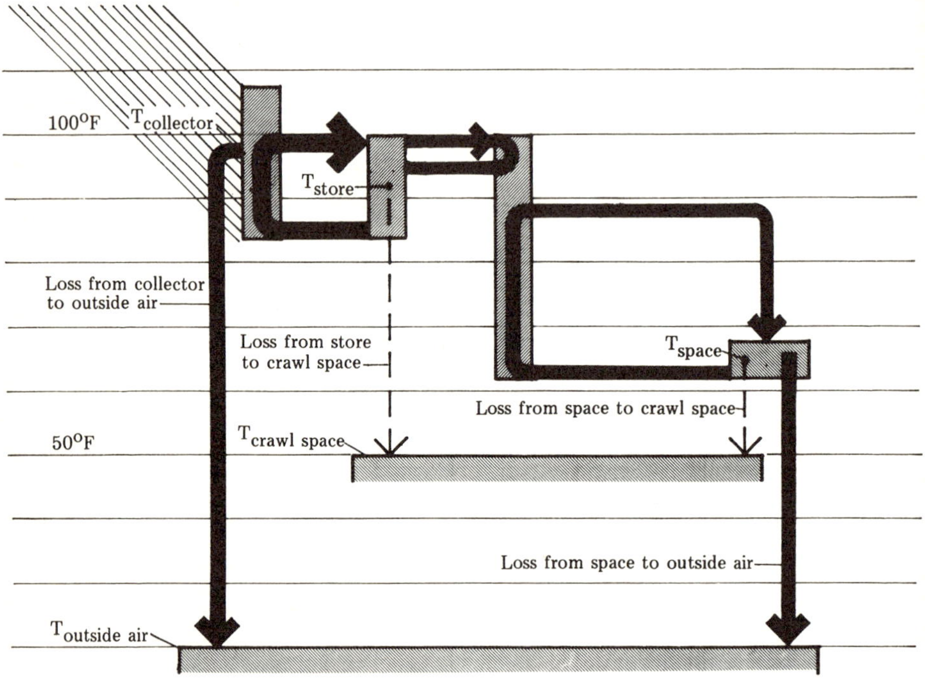

Fig. 1.4 Temperature schematic for typical daytime collection period.

store lost heat into the ground and into an insulated crawl space—some houses have stores which lose heat to the house itself, a complexity which can be ignored in the first instance.) Notice that the loss rate (represented by the thickness of the arrows) from the collector and from the house are very similar, while the loss rate from the store is much smaller. Also notice that the store loses heat to the ground and to the crawl space of the house, which are both about 55°F, while the collector and the house lose heat to the outside air, which in these instances is at a temperature near zero. Finally, notice that when the collector is off, no heat is lost, since its temperature quickly drops to that of the outside air (or a little below, due to nighttime radiation cooling).

Adding heat exchangers to the system merely increases the overall temperature drop from one end to the other, as shown in Fig. 1.6.

HOW THE SYSTEM COPES WITH SUN-PULSES

While it is common to think about solar heating systems as "getting hot enough" to heat a building, we propose that you clear your mind of such

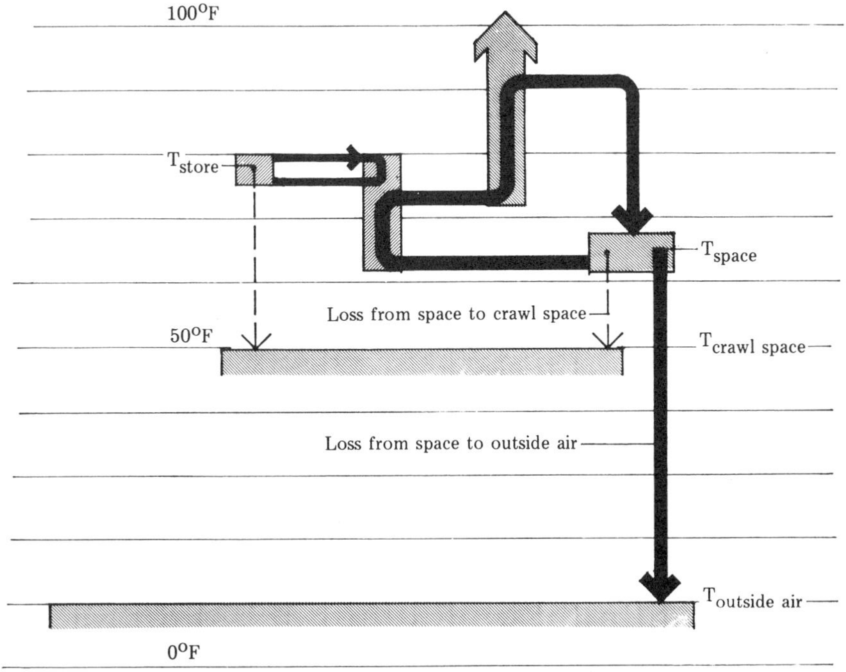

Fig. 1.5 Temperature schematic for typical nighttime discharge period. Note that the collector is turned off and is therefore outside the system.

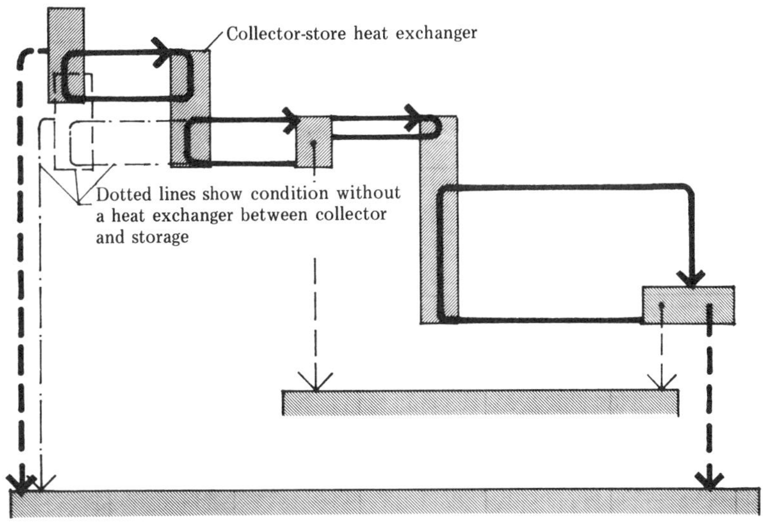

Fig. 1.6 Temperature schematic showing the increase in collector temperature, and resulting increased collector heat loss, caused by the addition of a collector-store heat exchanger.

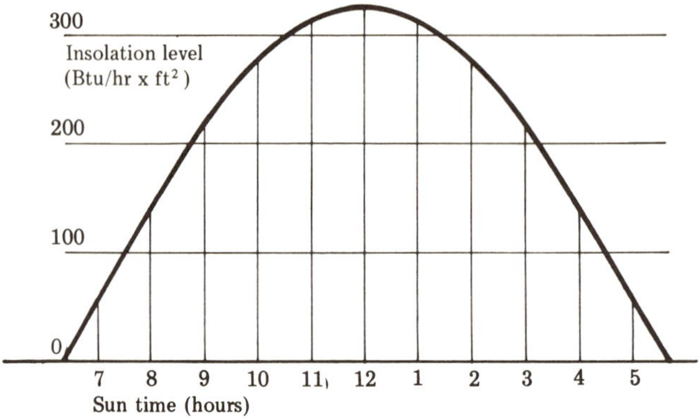

Fig. 1.7 Sun-pulse for a clear day, Latitude 40°N, on a surface tilted at 40° from the horizontal, on March 21st.

ideas and entertain a new way of thinking about such systems. If you have the good fortune not to have thought about these things before, you may find these ideas self-evident. But in any case, the mental exercise should be helpful and may even be fun.

While a building standing in the environment constantly loses heat, the sun delivers heat periodically, in "sun-pulses," a term we have coined to describe a graph of all-day insolation. These pulses, or at least the part of them which prove useful for solar heating purposes, occur in a wave shaped approximately like a half-sine wave, about 6 to 8 hours long (see Fig. 1.7). If the collector is large enough to do a noticeable amount of the space heating it will, during the middle hours of the pulse, deliver much more heat than the system can use. (As a rule of thumb, any system delivering more than 5% to 10% of the demand needs some kind of storage.) It is the job of the system as a whole, first, to collect as much of the available energy as possible, and, second, to lose as little of this heat as possible before it gets into the space or water to be heated. To accomplish this, the system must somehow soak up the intense sun-pulse and release it later to serve the "load" or heating demand. The essential issue in the system design is to accomplish these tasks while keeping the temperature of all parts of the system at the *lowest practical level* in order to minimize losses. If you look at the diagram of the system while the collector is operating (Fig. 1.4), you will notice that system temperatures vary from highest in one part of the collector (in fact, at the top corner of the fin, farthest from the water tube) to lowest in the building return air duct. The closer we get that top corner of the fin to the building return air temperature, the less we lose from the collector and the store, and the more we deliver to the house. It is all very simple in overall outline.

The importance of the delivery temperature in the design of the system cannot be overestimated. Many larger buildings, in order to cut down on distribution costs and simplify controls, distribute air or water at a constant temperature. Dispersed thermostats adjust the volume of flow in order to maintain the desired space temperatures. Compared with residential space heating, water heating, and ventilation loads, such commercial heating systems *must* operate at much higher temperatures, and so lead away from simple, low-cost solar collectors toward those which are less sensitive to high operating temperatures.

It is confusing that, in practice, most residential heating systems are operated at elevated temperatures, often above those found in commercial buildings. In a retrofit context, such elevated temperatures must usually be accepted; but in new construction, it is not only possible, but easy, to lower the actual operating temperature of the system to accommodate solar. Indeed, lower temperatures are the first step in energy conservation, and should for economic reasons be considered a necessity, regardless of solar.

A PREVIEW OF SYSTEM OPERATION

In our temperature diagram of the system, the various temperatures are carefully chosen, and their relationships are important. It is helpful to have at the beginning an idea of the overall behavior of the system as it goes through its various states during a year.

First, as the sun heats up the collector plate, the plate's temperature rises, until fluid is passed through it to begin the cycle. The entering fluid, drawn from the bottom of the storage tank, is the coolest element in the working part of the collector, and it emerges at a temperature below the highest temperature in the collector. But the two ranges overlap considerably, a way of saying that the collector is an effective heat exchanger (see Chapter Five). Next, the fluid returns to the tank which, in a well-designed water system, is almost the same temperature at its top as at its bottom. The fluid begins to heat up the top of the tank, increasing its temperature stratification (difference from top to bottom), of course increasing the stored energy (since the water is getting hotter in the collector), and raising the average temperature of the store.

This process continues through the day, until toward the end of collection the temperature rise of the tank levels out, the stratification decreases somewhat, and finally fluid stops flowing through the collector when the pump is turned off.

On the other end, a pump circulates water from the top of the tank to a heat exchanger in the building's return airstream. Unless the thermostat setting is changed, the space stays at a roughly constant temperature throughout the day. The amount of heat flowing from the storage tank into the space varies continually, however, as the temperature at the top of the store rises, and as the thermostat regulates the supply of heat to the space. Assuming it is cold enough outside to call for heat all day, the rate at which the store discharges into the space does not depend upon how cold it is outside, but only on the difference between the temperatures at the top of the store and that in the return air passing over the coil.

After the collector is turned off, the heat exchanger continues to run, gradually lowering the temperature of the store and further reducing its stratification until, by the next morning, the store is roughly in the state it was the morning before.

In warmer weather (when space heating is not needed continuously) and on the second of two successive sunny days (all other things being equal), the system will operate at a higher temperature than in cold weather or on days of alternating sun and clouds. The sequence of sun and cloud makes a difference to the system (although exactly what difference remains to be seen), as does the sequence of warm and cold days. A good system simulation must take account of this by "exercising" a system through a complete model year of sun-pulses and temperatures.

There are no fixed values for the capacity of the store, the output of a system on a day, the length of "carryover," the time it takes to complete a cycle, etc. A well-designed system is totally dynamic.

2

COLLECTOR LOCATION

SUN ANGLES

There is no particular reason why a designer of a solar system needs to know the astronomical relations between the sun and the earth. But, if these relationships *are* understood, they can be of use on occasion when tables are absent and a quick assessment needs to be made. Students should know these things as a matter of general education. After all, the lowliest of shepherds knew much more of what was going on in the heavens than the average college graduate of today. If we are to rely more heavily upon ourselves in the future, there is no better place to start than to recover this lost body of general understanding.

Our purpose is to track the relationship between the direct beam of the sun and a fixed point on the earth. By imagining how the earth looks from the sun as in Fig. 2.1, it is possible to trace the point on the earth where the sun is directly overhead, and to generate a series of lines on the earth's surface which records various limits to the sun's direct beam.

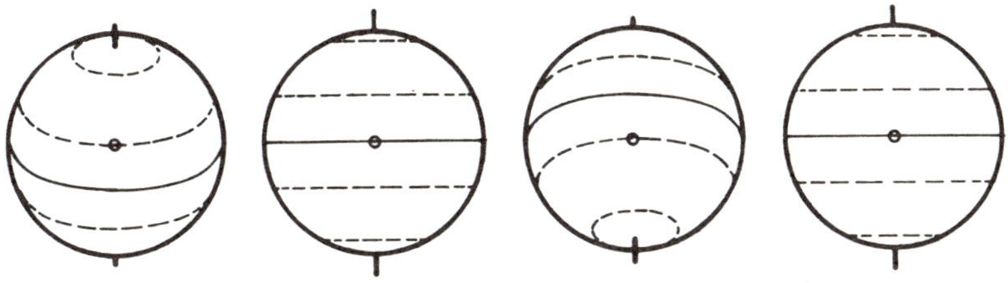

Fig. 2.1 The earth as seen from the sun in four seasons. The spot represents the point on the earth where the sun appears directly overhead. (From diagram by the author in "Solar Dwelling Design Concepts," HUD-PDR-156(2)).

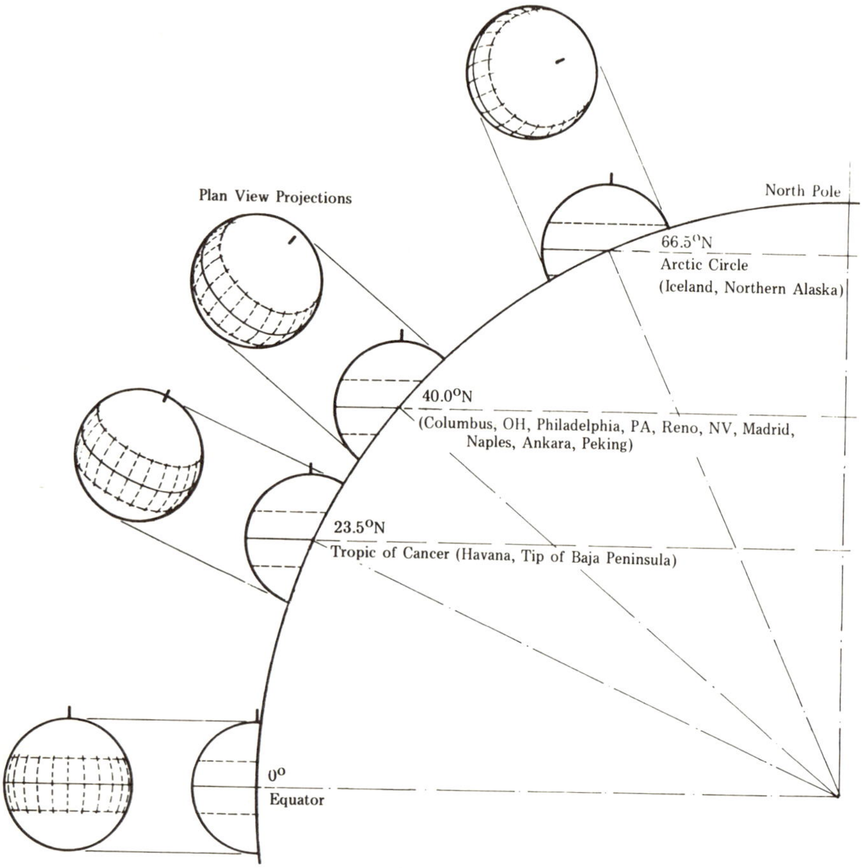

Fig. 2.2 Section through the earth, showing the side and plans views of the sky dome at various latitudes on the earth.

The tropics, for instance, record the latitude beyond which the sun is never directly overhead; the circles, those latitudes above which the sun is both up all night and set all day one or more days of the year. The extremes of the earth's apparent "wobble" north and south during the year (as viewed from the sun) occur on the solstices; the midpoint, when the sun stands over the equator all day, occurs on the equinoxes.

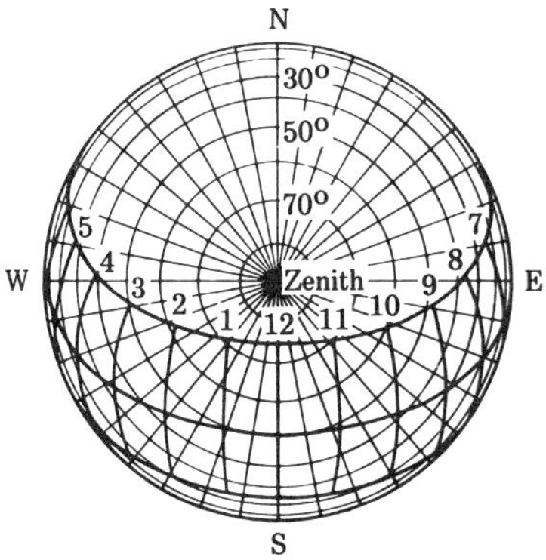

Fig. 2.3 True plan projection (as shown in Figure 2.2) of the sky dome at Latitude 40°N, showing sun's path at the equinoxes and solstices, and showing the hourly location of the sun.

Now imagine that a transparent dome (we will call it the sky dome) is placed with a fixed viewing position for your head at its center. If you were to trace onto the dome the position of the sun as it crossed the sky on one equinox and both solstices, what would the sky dome look like when you had finished? The answer, as shown in Fig. 2.2, is that it would look exactly like a conventional earth globe with an equator (the sun's path during the equinoxes) and the two tropics (the sun's path during the solstices). The sky dome's new north pole would point in the same direction as that of the earth itself. This result would obtain anywhere on the earth's surface, as shown in the figure.

Fig. 2.3 shows a plan view of the sky dome for latitude 40° N, the sun paths show up as ellipses—drawing such a plan is easy if you know simple projective geometry. The position of the sun hourly is found simply by drawing lines of longitude 15 degrees apart, because it takes the sun 24 hours to cross 360 degrees of longitude, and $360/24 = 15$.

We on the surface measure things from our own reference system. Our zenith is straight up, our horizon is horizontal, and the longitude lines of our own sky dome radiate evenly from the zenith. The diagram shows that the sun follows simple paths relative to a similar coordinate system *tilted* from our zenith by the latitude of our location. We obviously need a *map* which clearly translates the coordinates of the sun's path in the sky into those of our own sky dome. The most common one, such as that in Fig. 2.4, places the local zenith at the center, with longitude lines evenly spaced radii, and latitude lines spaced evenly along the longitude lines. It is as though the plan view of the sky dome were stretched at the horizon until the latitude lines no longer bunched

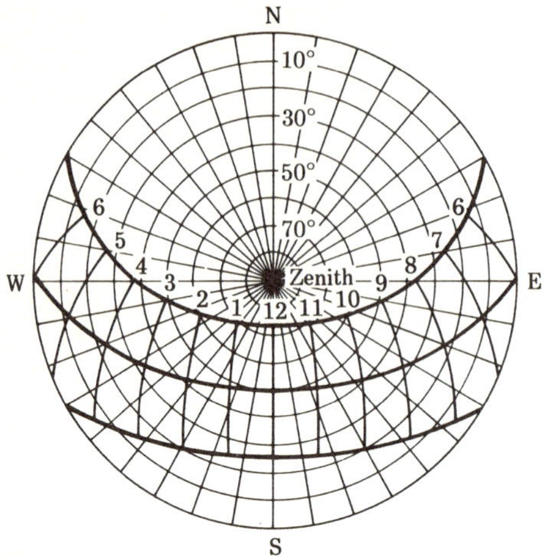

Fig. 2.4 Map of the sky dome developed from the true plan projection in Fig. 2.3 by extending the lines of latitude to form equal intervals from the zenith. This is a simplified example of the charts found in a number of references, such as "Time Saver Standards," published by McGraw-Hill. Notice the apparent distortion of the sun's paths, which show as regular ellipses in the true plan in Fig. 2.3.

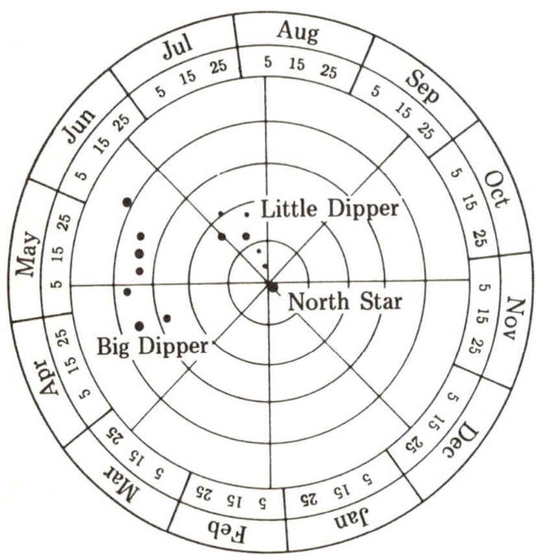

Fig. 2.5 Sky map showing the position of the big and little dipper and Polaris (the north star) throughout the year, at 8 PM. Hold the chart with the date at the top, and face north. The two outer stars of the big dipper bowl point to Polaris, which is very close to the celestial north pole.

up around the outside edge. Because of this distortion, the paths of the sun, when plotted on such a map, look distorted. But we now can read the numbers clearly, making the map an invaluable tool for sun watchers.

When checking a site to see if there is enough sun for a solar heating system to be practical, it helps to have a tool at hand and a few numbers in your mind. The tool you need is a compass—remember that compasses are sensitive to metal, and don't read accurately when too close to steel or reinforced concrete. You must know your local magnetic declination, to find true north from compass north. (If you can see the site on a clear night, set your compass by the North Star and avoid any possibility of misreading.)

The numbers you need are the *cotangent* of the sun's angle above the horizon for a few key dates—usually the winter solstice and the

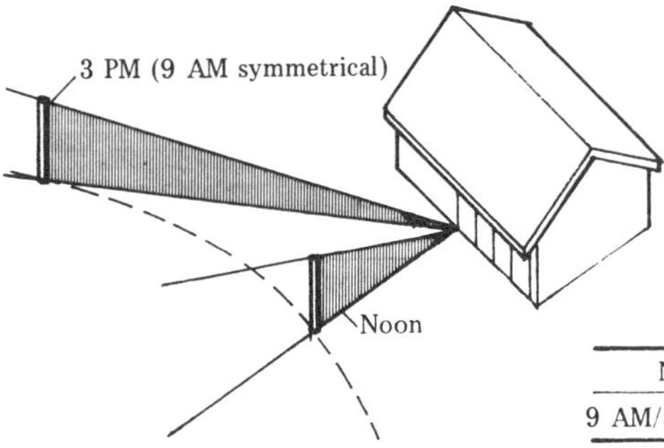

	25°	30°	35°	40°	45°	50°
Noon	1.13	1.35	1.63	2.00	2.54	3.37
9 AM/3 PM	2.16	2.57	3.14	4.00	6.42	8.86

Fig. 2.6 Table showing the length of the shadow from a pole of unit height at various latitudes, for December 21st, at 9 AM/3 PM and noon. The numbers also represent the cotangent of the elevation angle of the sun. In all the cases shown on the chart, the sun is approximately 45° from true south at 9 AM and 3 PM.

equinoxes will do. If you remember the cotangent of the angle and not the angle itself, you will never be caught short if you don't have a protractor or sun-angle device (of which there are several on the market). Fig. 2.6 will help you work out these numbers. The sun's angle at 9 and 3 is used because those hours bracket the usual limits of useful collection at the winter solstice, which is the worst condition.

It is useful to remember where the sun rises and sets. At the equinoxes, the sun rises everywhere due east at 6 A.M. sun time, and sets due west at 6 P.M. sun time. At the solstices, the sun rises and sets symmetrically north or south of due east, the angle depending upon the latitude. At the equator, the angle is the same as the tilt of the earth—23.5 degrees. At latitude 42°, the angle is about 30°; that is, the sun rises at azimuths 60° on June 21 and 120° on December 21. It is worth noting that the sun shines out of the north half of the sky at least part of the day for six months of the year everywhere on earth, which explains why it is hard to pick up summer sun in the early morning and late afternoon.

If you are on a site without reference tables, it is tempting to estimate the sun's angle from true south as it moves through the sky. Remember that the sun moves 15° per hour across the sky in the coordinate system of the *tilted* sky dome, and not in the local coordinate system. For instance, at latitude 30° north, on June 21 (summer solstice), the sun traverses 70° of sky longitude between 11 A.M. and noon.

These astronomical considerations define the most important element in solar collection: does the sun shine directly on the collector? From our everyday experience, we know that the direct beam of the sun delivers the major portion of the energy; granted, as we will see, some energy comes in from the rest of the sky dome. But if adequate direct sun can strike the collector, enough of this other, "diffuse" radiation will also strike it. We still know nothing about the *intensity* of this beam of radiation and of the diffuse radiation, both of which depend upon issues to be discussed in subsequent sections.

Little has been said about shading, and nothing about the tilt and orientation of the collector itself. While these issues are usually lumped together with astronomy, they are really very different. Tilt, orientation, and shading, are *economic* matters, involving trade-offs in the design of the structure, the solar system, and the community. There is nothing we can do (thank heavens and pardon the pun) about the relation between the sun and the earth; but we *can* decide not to solar heat a building which points the wrong way. How we make such trade-offs is the subject of the next section.

SHADING AND SITE PLANNING ISSUES

In the jargon of solar energy, "penetration" refers to the extent by which solar displaces other forms of energy. It is a useful concept because it can be applied to any system at all. One can define the solar penetration of a home heating load, or the overall solar penetration of an electric utility. What does penetration have to do with astronomy?

Many studies have tried to predict the future penetration of solar into various energy sectors—residential, commercial, industrial, and transportation being common examples. These studies typically arrive at percentages based upon careful projections of many social, technical, and economic factors into the murky future. Little effort is made to graph the relationship between penetration and other factors. The interrelation of site planning, building design, astronomy, and penetration is hereby proposed as the basis for such a graph.

First, consider a particular building. By defining the aperture of the proposed solar collector (passive and/or active), the well-known astronomical mathematics outlined above can be applied in order to define the outline of a shaft in space which must be reserved, clear of interference, in order to insolate the building's collector(s). Right away a question arises: do we include every ray of sun on every square foot of collector? This would require that the "space reservation" extend to the horizon, obviously not a practical idea. As we will see further on in Chapter 4 collector output falls off dramatically as the angle of the sun becomes more nearly parallel to the surface (as the "angle of incidence," that is, the angle between the sun ray and a perpendicular to the surface, increases, to put it another way.) We can begin to simplify the problem by excluding rays which exceed a 60° or 70° angle of incidence.

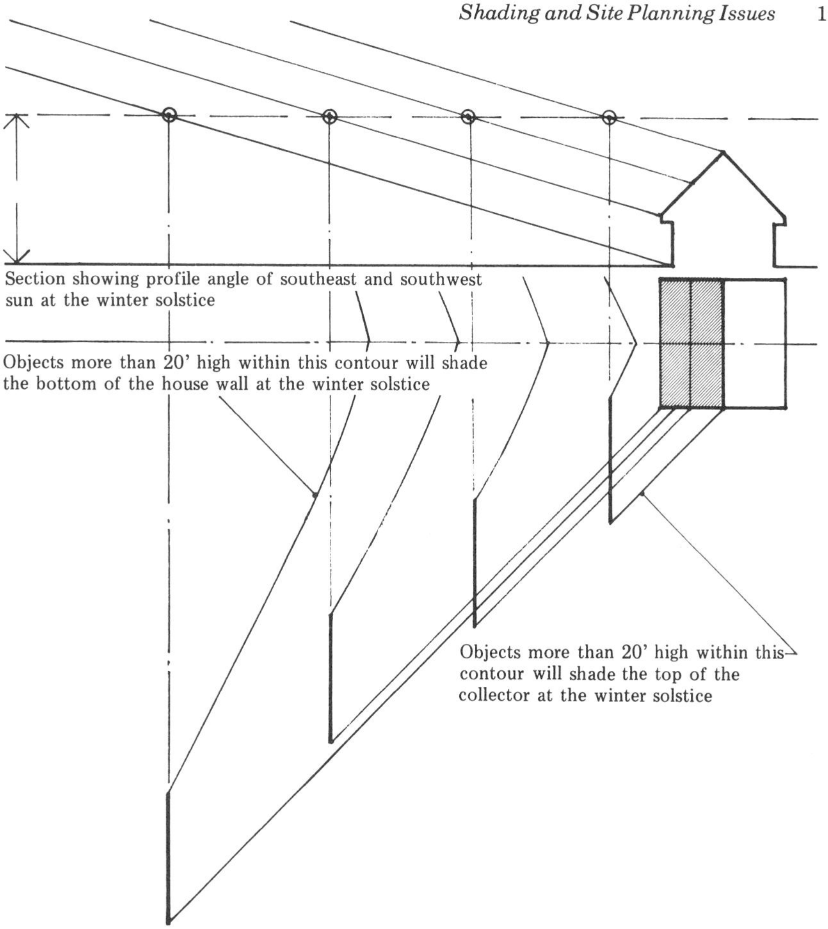

Fig. 2.7 Solar interference contours define the area within which objects which rise above a given height relative to a collector will shade the collector for a given date. Because of the large number of such charts necessary to define the required space reservation fully, most solar planning techniques use the shadow cast by an object, which resembles a solar interference contour in reverse. Shadow studies, however, tend to neglect the shortening of the shadow as it falls on elevated objects such as roof-top collectors, and as it falls on sloping ground.

Even with this simplification, the space reservation is very large, as shown in Fig. 2.7 for a typical building at latitude 40 degrees north. Of particular concern is the distance away a similar building must stand to avoid interference. This is directly related to population density, which is a closely measured and important characteristic of human settlement. Certain factors relate density and the amount of available sunlight:

- the shape and height of the buildings—detached houses can be packed closer together if they are wide and low than if they are narrow and tall, given the same solar space reservation;
- the latitude—the higher the latitude, the lower the density;
- the time of year when energy is needed. If energy is needed only in the summer, when the sun is highest in the sky, the adjacent buildings can be much closer.

Starting with the largest reasonable space reservation for a particular solar building, one can begin to trim its size in various ways. First, the lowest sun can be excluded, and the earliest and latest hours. Then, a certain amount of partial shading can be allowed, caused perhaps by building elements or trees. The collector can be made smaller. The load can be shifted from space heating, which requires a large collector used in winter, to hot water heating, which requires a small collector used in summer. As each successive cut is made, one of three things happens:

- the system is made larger, to produce the same output as before, given the reduced input;
- the system remains the same, but the output goes down (as does the system's cost effectiveness);
- the system is made smaller, maintaining its cost effectiveness, but reducing its output as in the second case.

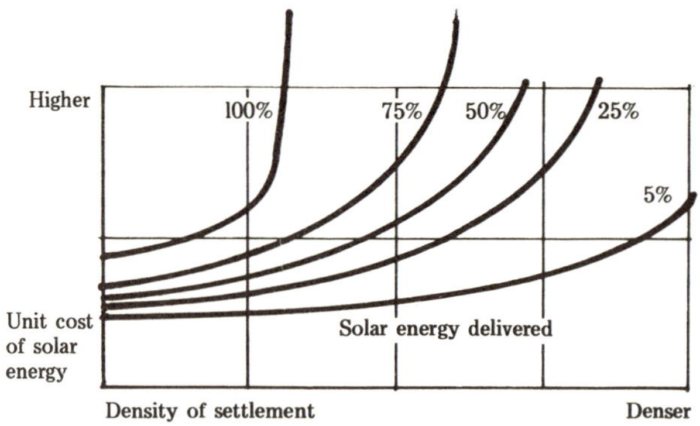

Fig. 2.8 Hypothetical graph of the relationship among settlement density, unit cost of solar energy collection, and the solar fraction. Such a graph would apply to an individual building or development, or to an entire community. If a real graph of this type could be developed, it would be a useful planning tool, because it would indicate the reasonable potential for on-site versus off-site solar energy devices.

For the building, considered as a system, the solar penetration level is a function of the extent of the space reservation (the smaller the reservation, the lower the penetration) and of cost (an increase in cost can in many cases compensate for a reduction in supply). A graph could be made, relating the penetration level with cost and density, and it might look something like Fig. 2.8 in a typical case. Since this is a typical case, it exemplifies the behavior of a whole community, with regard to on-site collectors. When density gets too high, most of the buildings are in shade most of the time (Manhattan being the almost ridiculous extreme), and off-site solutions are required. In fact, as the cost per unit of solar on-site gets higher, a point will be reached sooner or later where off-site solutions become cheaper.

These matters lead toward national policy and, while interesting, do not concern us here. With regard to a single site, what matters is that shading is not a technical concern but a social and economic one. The building is exposed to a pulse of sunlight, of a shape determined in part by the extraterrestrial insolation, in part by the weather, and in part by the surroundings. The first two are in the realm of physics, the latter of zoning ordinances, land costs, easements, law, friendly neighbors, developers—in short, all those complex issues which make up the texture of everyday life.

But the sun striking the building or the site is one thing, and the sun penetrating the collector cover is another. What if, again for siting and economic reasons, the collector does not have the good fortune to face due south at the optimum tilt angle? This time there are only two choices:

- the collector can be made larger, to compensate for the unfavorable angles, at an increase in cost for the same output;
- the collector can be left the same, with a decrease in output, and a decrease in cost effectiveness.

There is, unfortunately, no way to restore the original cost effectiveness by reducing the size of the system. If the sun strikes a large system obliquely, it will strike a small one so. At some point, again, an off-site solution is the cheaper.

One important class of economic trade-off involves collectors arranged in parallel banks (or arrays—there must be a difference between the two terms, but it has escaped the author). The term "profile angle" is useful in defining the problem. It is the angle made by the sun's direct beam projected onto a plane perpendicular to the collectors. Assuming there are no obstructions to the southeast and southwest (there usually are), the profile angle can be used directly to optimize the spacing of parallel banks of collectors, as shown in Fig. 2.9.

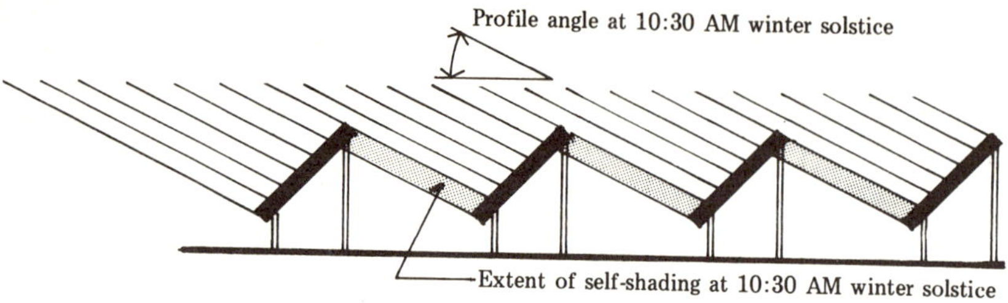

Fig. 2.9 Profile angles can be used to determine the extent of self-shading in parallel collector banks. In the example shown, no shading occurs between 10:30 AM and 1:30 PM solar time at the winter solstice (Dec. 21st).

Briefly, the problem involves accepting a certain amount of self-shading in order to increase the density of collectors on the roof and avoid long piping runs. The problem is quite similar to that encountered on a larger scale in site planning, except that in the present case all the elements of the cost trade-off are under the control of a single owner. What happens when the lower part of a collector is shaded? The effect is identical to reducing the sun's intensity on a fully lit collector by the percent of shading. As we will see, the effect can be modeled perfectly by using "sun-pulses" with a shorter time of day. By modeling the percent reduction over typical days throughout the year, the overall reduction in collection can be balanced against the added costs involved in spacing the collectors further apart.

Summarizing: shading, orientation, solar access, and collector tilt are all social and economic issues. The sun, if obliged by the weather, sends its beneficial rays toward the site. Whether these rays are collected depends upon the pattern of settlement. If we want a maximum of solar collection on-site, we must arrange our buildings accordingly. If not, we must pay progressively more for the solar we collect; or collect progressively less, and supplement the on-site supply with energy from off-site sources.

Fig. 2.10 gives some idea of the economic consequences of various suboptimum siting arrangements. These must be taken account of in any solar design. Ignoring these factors is common in the design of subdivisions and communities, and is not likely to decrease as companies experience the competitive pressures to "solarize" existing homes. We may perhaps be spared north-facing collectors, but every other economic catastrophe is sure to be perpetrated somewhere through ignorance or guile.

Fig. 2.10 Two hypothetical cul-de-sac site plans, showing how careful placement of buildings and trees can vastly improve the potential solar heating fraction. The difficulty of accomplishing a change such as that shown increases with density.

GUIDELINES FOR DESIGNERS

The first step in assessing a building for potential solar heating is to locate the collectors. More precisely, a plane or series of planes must be found with these characteristics:

- an adequate amount of sunlight strikes it at the right time
- the sunlight is reasonably secure
- active or passive collectors can be mounted in these planes without intolerable insult to the architecture
- the resulting cost will be acceptable.

It is impossible to provide exact trade-offs, especially with regard to retrofit installations. In practice, an owner needs to know roughly whether solar heating is feasible *in his terms*. It is the latter element which makes exact trade-offs almost useless: a good salesman can persuade an owner to choose a larger system than can someone more cautious. Few owners choose solar solely on the basis of cost effectiveness, and even those who do have widely varying definitions of what is cost effective. What is needed at the initial stages of establishing feasibility is a general sense of where to put collectors, what kind (active or passive) to use, and whether the installation is ideal or marginal. The following checklists are intended to satisfy this need.

RETROFIT INSTALLATION CHECKLIST FOR SHADING AND COLLECTOR LOCATION

Find True South

- Know the local compass declination (the angle between compass north and true north). If in doubt, set the compass relative to a known direction, such as a true-north street, Polaris, etc.
- Watch out for compass deviations caused by reinforcing steel, large steel objects, etc.
- Most maps show *compass* north, and fail to include the declination. Also, declination drifts a little over time, and old maps may not correspond even to compass north.
- In his office, the author has assembled a collage of Geological Survey maps for his region (the Boston area). Outside densely populated areas, the maps often accurately show individual houses; and streets, if not too winding, are shown in their true orientation. It is often possible to establish the orientation of a house over the telephone, while looking at the map, and incidentally it aids in finding the client's house.

- Architects and engineers will often use a "plan north" parallel to the building, with an unknown orientation (sometimes more east or west than north). Don't rely upon arrows drawn upon an architectural plan—only on civil engineering site plans.

Find the Most Likely Location(s) for the Collectors

- After a little practice, it is possible to "eyeball" acceptable shading conditions, and determine the only reasonable location for collectors.
- If collectors are to be mounted on a flat roof, it will almost always be necessary to perform a structural analysis, to see if the roof can support the dead load and, most important, resist the wind load. Many commercial and industrial buildings are constructed with very little residual strength in the roof structure. In the author's experience, this factor limits flat-roof retrofit installations severely, and often requires expensive, long-span support structures.
- It is ideal to mount active collectors directly on a sloping roof, perhaps on simple stand-offs, since the installation is cheaper and more attractive than that requiring a support structure. On the other hand, it is often cost effective for small installations to mount the collectors on pipe frames at an ideal tilt and orientation. The trade-off usually occurs at about 35° off-south orientation; and, in snowy climates, at tilts below which the snow will not easily avalanche off the collectors (around 35°).
- Always consider the possibility of a solar attic, whether active or passive. Despite the problems of glazing a roof, the advantages are great: better appearance, creation of a nice space (if passive), and good cost effectiveness if the roof needs replacing.
- If collectors have to be ground mounted, it is usually necessary to use antifreeze, because the piping will not be able to pitch uniformly to storage. Except in warmer climates with dry earth, avoid buried storage tanks. They are expensive, hard to get at, and all heat lost from the tank is useless. Also, buried and insulated piping is incredibly expensive. For instance, 150 feet or so of double buried piping might double the cost of a water heating installation.

Observe Shading Conditions

- Using the chart of tangents and azimuths at the beginning, end, and middle of the sun path for the winter solstice (Fig. 2.6), determine the extent of midwinter shading at the bottom and at the middle of the collector.

- Since collectors will often be mounted high on the building in retrofit installations, it is sometimes necessary to stand to the north of the building in order to sight through the collector; or a ladder or high window can be used to get up where the collector is mounted. Because of the difficulty of such gymnastics and the short time usually available for the feasibility study, the designer needs quickly to develop an eyeball sense for shading. Make sure this sense is checked periodically against real measurements!
- Various devices are available or possible for determining shading with some accuracy. Remember, however, that no device can integrate the shading over the entire collector for the entire year. Typically, measurements have to be made at various points on the collector.
- Deciduous trees pose a problem. Some, like certain oak species and beeches, do not lose their leaves in winter; and others have dense branching. Even if lightly branched, several layers of tree can produce a dense shadow. Remember that the loss rate of the collector does not change because of shading, only the gain. For cold conditions, a substantial reduction in gain might make the difference between a net gain and a net loss. Even if the shaded part of a collector array is left unused, thus eliminating the loss, it represents a capital investment not amortized.
- Establish the likelihood of continued solar access with some assurance. Absolute certainty is usually impossible and unnecessary. If future shading is expected, design as if it were there now.
- Shading is not as serious for water heating as for space heating, since much of the load occurs during high sun angles. If midwinter shading occurs, decrease the collector tilt.

NEW INSTALLATION CHECKLIST FOR SHADING AND COLLECTOR LOCATION

(Read the Retrofit Checklist First!)

Find True South

- Visit the site and treat the problem like a retrofit; or use maps with known compass or true north bearings.

Site and Design the Building to Control Solar Gain

- Custom buildings should be designed to suit a particular site, with careful trade-offs made from the outset among plan, building form, access, parking, views, solar gain, location on the site, etc. In the case of a custom house, if such trade-offs are not made properly, the owner will be after the architect's scalp in a few years when he finds out what he missed. If solar gain is limited this must be clearly demonstrated to the owner.
- Builder's houses typically are designed without reference to a particular site. It is the height of folly to maximize the passive and/or active potential of a builder's house, and then to site it just north of a stand of growing evergreens. However, if all other things are equal, and south windows are maximized at the expense of north ones, even a small amount of gain will make the change in plan worthwhile. Advertising such a house as a solar house is another matter.
- Prevention of unwanted solar gain in summer is crucial in almost any climate, and is especially important where there is a large cooling load, and in nonresidential buildings which have a large internal heat gain. Ideal solar heating orientation might well be sacrificed for better summertime gain control—for instance, facing the long side of the building to the southeast instead of south.
- Even for passive buildings, the provision of south-facing sloping roofs or provisions for future collector racks is a good idea, assuming that photovoltaic cells continue their rapid decline in price. Garages and outbuildings can also be retrofitted later for PV cells, but at some cost penalty due to the additional cabling required.
- Take account of east and west gain, since during spring and fall, both will help serve the heating load. But be very careful of the summertime gain problem. It is very difficult to shade east and west facing windows, since the sun will be shining directly in at low angles during hot weather.

Shading Considerations

- Avoid placing collectors in notches or next to projecting wings which will cast shadows on the collectors during prime collecting hours.
- Avoid projections below the collectors which will retain snow and cause self-shading. Snow will avalanche off collectors nicely if it has some place to fall. Even a slightly projecting gutter or glazing

mullion will catch snow and inhibit prompt avalanching. Do *not* try to melt any substantial amount of snow off the collector, as it would probably eat up a year's energy harvest!

- Try to plant low-growing trees or shrubs to the south of the collector (whether active or passive). These will also provide a better visual barrier and windbreak than tall trees with bare lower trunks. In humid climates, however, planting might best be taller to allow cooling breezes to flow beneath the foliage and to shade the building. Again, solar heating may have to give way to proper cooling measures.
- As with retrofit installations, establish at least some assurance that continued solar access will be available.
- With parallel banks of collectors (Fig. 2.9), trade off the dimension between collectors against self-shading. For space heating, avoiding self-shading between 10:30 and 2:30 sun time at the winter solstice is a good conservative measure. Water heating systems can afford more self-shading, especially in very cold climates where wintertime collection is minimal. Sophisticated procedures have been developed to predict the effect of easily calculated self-shading, but the author suspects that such procedures seldom include all the relevant factors involved in cost effectiveness, and prefers simple rules of thumb.

3

THE SUN-PULSE

THE CLEAR-DAY TRUE-SOUTH SUN-PULSE

A pyranometer is a small solar collector which picks up and integrates the direct and diffuse radiation coming from the hemisphere it "sees" (see Fig. 3.1). Because its glass cover is hemispherical, and its absorber very small, all the radiation comes in normal to the glass, eliminating off-angle reflection. (See Chapter 4.)

If a pyranometer is placed at the same tilt and orientation as a solar collector, it will receive and record the rate of solar power impinging on the collector itself, at any given moment. Suppose we plot the variation of this power on an hourly basis, throughout a relatively clear day. Such a curve for the Acorn Solar Test House in Acton, Massachusetts is shown in Fig. 3.2.

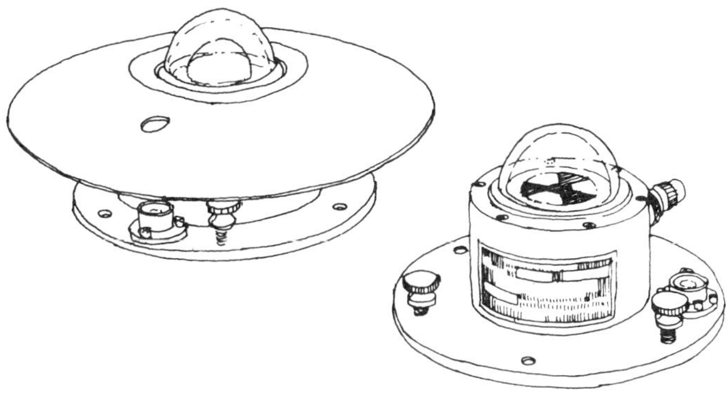

Fig. 3.1 Sketches of two typical precision pyranometers. The larger one is a Class One device, the smaller, Class Two.

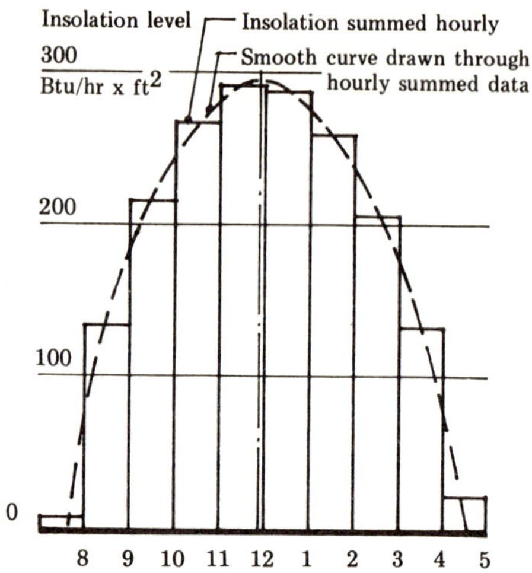

Fig. 3.2 Graph of hourly summed insolation data derived from a Class Two Eppley pyranometer tilted at 47° from the horizontal, at the Acorn Test House in Acton, MA, on a clear day in January, 1976. The smooth curve represents a much more accurate graph of the instantaneous insolation level. The area under each curve is equal.

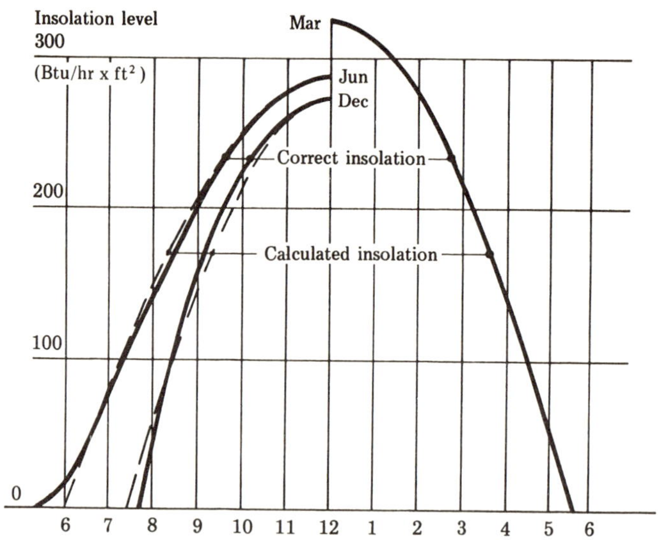

Fig. 3.3 Graph of clear-day sun-pulses for a flat-plate collector at latitude 40° north, tilted 40° from the horizontal, derived from published data (ASHRAE Publication 170); and superimposed cosine curves with equal areas under the curves. Note that the largest errors occur at sunrise and sunset, when collectors are normally not functioning; and that the errors are very small or non-existent (in March).

Note that the recorded data sums the energy received during the preceding hour. By drawing a smooth curve through what is really an hourly bar graph, one appears to be creating a refined and precise graph of the instant-by-instant power level. This can be misleading, a point that will be discussed later in this chapter. Keeping in mind its illusive accuracy, one can closely approximate the curve by a half-sine wave made to fit the curve empirically.

Fig. 3.3 shows the clear-day sun-pulse curve for three months, for a collector tilted at 40° from the horizontal, at 40° N latitude. It can be seen that, especially in the summer, the curve deviates a little from a simple half-sine or half-cosine curve, but can be modeled by such a curve with good accuracy.

If it were possible to assign once and for all the appropriate sine-wave-shaped sun-pulse to a given place for a given month, the advantages would be enormous. First, the area under the curve and the value of the insolation at any moment could be quickly and automatically calculated. Second, the sun-pulse can be manipulated in many ways to mimic the effect of various kinds of weather and atmospheric conditions. Third, the sun-pulse can be derated to account for off-angle reflections, transmission losses and absorption losses in the collector, dirt, and shading. By classifying and standardizing these manipulations, the effects of virtually any weather pattern can be modeled using easily stored numbers. By associating sun-pulses with appropriate values for ambient temperature and wind (which in their turn can be reduced to simple patterns), all the variables which influence the heat gain and heat loss for both a collector and a building can be modeled with simple equations.

DERIVING THE BASIC SUN-PULSE DATA

When the earth is its average distance away from the sun, about 429 Btu/hr × sq ft of solar radiation reaches the outer atmosphere. This quantity, known as the "solar constant," had been overestimated by about 3%, until satellite measurements were made. It measures the average of a quantity known as "extraterrestrial radiation." The average value of 429 Btu/hr × sq ft occurs in April and October; the minimum is about 415 Btu/hr × sq ft in July, while the maximum of about 444 Btu/hr × sq ft occurs in January, a variation of about 3.5% above and below the average. One can surmise that the northern hemisphere is more comfortable than the southern, because the greatest intensity of sunlight occurs during the winter and the least in the summer, while the reverse is true south of the equator.

The spectral distribution of this radiation is close to the theoretical radiation from a "black body" or perfect radiator, at a temperature of

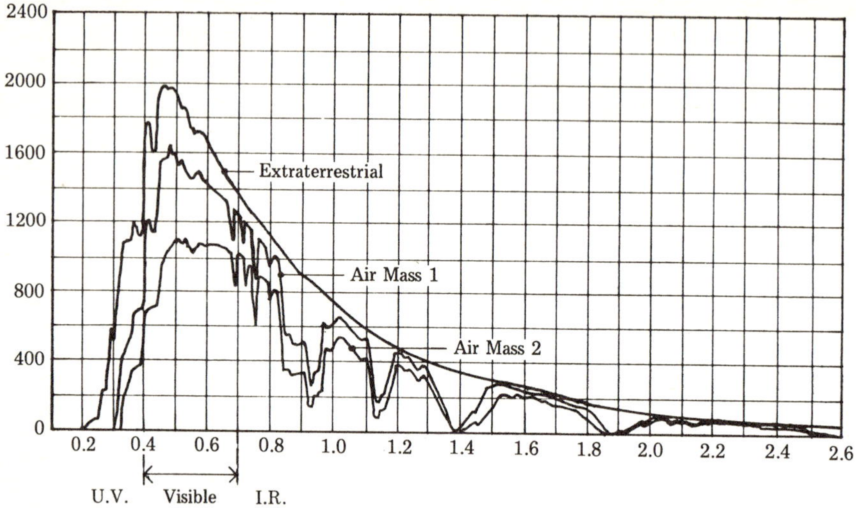

Fig. 3.4 Solar spectra with the earth at the mean earth-sun distance of the direct beam radiation through a clear sky at sea level, for air mass one (middle curve) and air mass two (lower curve). The top curve represents the extraterrestrial radiation under the same conditions. The data is derived graphically from a number of standard texts, and is intended only to show the general character of each curve.

9912°F (5762°K), except that relative to the black body curve more visible light and less UV is emitted by the sun. Of greater interest to the solar designer is the state of the radiation after it passes through the earth's atmosphere. Fig. 3.4 shows the spectrum before and after it passes through one and two "air-masses," a measure of the thickness of a standard, clear sky. If the direct beam of the sun strikes the ground at an elevation angle of 30°, it has passed through two air masses (assuming it would have passed through one if it shined from directly overhead). Obviously, as you move up in elevation, the air mass through which the sun must pass decreases, allowing insolation rates in mile-high cities to climb 10% to 15% higher than those at sea-level. Very dry and clean air from the arctic will also reduce the effective air mass in the winter in the northern parts of the country.

In passing through one air mass, the direct beam is reduced from 429 Btu/hr × sq ft to about 307 Btu/hr × sq ft, a drop of about 28%. This attenuation is caused by several factors: scattering of energy back into space by molecules in the atmosphere; scattering of energy toward the earth from all directions, producing some of the "diffuse" component of sky radiation and causing the sky to appear blue; and absorption of energy, which is reradiated as diffuse radiation, by the ozone, water vapor, and carbon dioxide which comprise part of the atmosphere. Generally, the molecular scattering reduces the outline of the spectrum, while molecular absorption produces the sharp gashes in the spectrum.

Although complex in itself, the calculation of the direct beam radiation under clear conditions is relatively straightforward, and in fact has become the backbone of the various scientific efforts to rehabilitate the data which has accumulated over the years in the files of our poorly-funded National Weather Service data-collection programs. But one more element (which has not been a major problem) must be mentioned, namely the gradual deterioration of the atmosphere in urban areas due to industrial pollution. This adds "turbidity" or opaqueness to the atmosphere by introducing many new chemicals, each of which absorb part of the incoming radiation. Much of the absorbed radiation is reradiated in the form of diffuse radiation, but some is lost to space. Even new weather data, such as the rehabilitated "TMY" (Typical Meteorological Year) data, is based partly upon weather records taken in the 50's and 60's, before many areas experienced extensive atmospheric pollution. This factor must be taken into account by the designer in reading weather data.

The major complexity in computing sun-pulses is finding the distribution of direct and diffuse radiation on days which are not clear. This would be of little interest to us if we had data which had been collected by tilted or vertical instruments. Because insolation data was primarily of interest to farmers until recently, it was taken almost universally in the horizontal plane, by instruments such as the pyranometer shown in Fig. 3.1. The only output is a record of *total* insolation striking the horizontal plane. In most cases, the instrument was placed out in the open, so that it "saw" the entire sky dome from horizon to horizon. To tilt the data, we need to know how much is direct and how much diffuse. The direct component is tilted by a geometric formula which finds the cosine of the angle of incidence between the tilted plane and the direct beam; while the diffuse component is tilted by cutting off that part of the sky dome which is now invisible to the tilted plane. This latter calculation makes the assumption that the diffuse radiation comes in evenly from all directions, which is clearly false in some cases—for instance, when clouds cover the sun and one quadrant of the sky vault, while blue sky shows in the other quadrants. However, no practical alternative has been proposed.

A good deal of research has been done on the average distribution of direct and diffuse radiation. It has been correlated reasonably closely with the extent amount of cloudiness, which is expressed by the set of terms $\overline{K}_T$, K_T, and k_t, depending upon whether the averages are monthly, daily, or hourly. For the hourly case, one correlation says that on the clearest day, 15% of the radiation will be diffuse; and that as the cloudiness increases, the ratio of diffuse to direct also increases, until at about 40% of a clear day, all the radiation can be considered to be diffuse. The actual correlations are related back to extraterrestrial radiation values as a baseline.

Obviously, such correlations cannot be considered to define every real hour. For instance, consider an hour consisting of one-half hour of full, brilliant sun, and one-half hour of dense fog. The response of a collector to that hour, compared to one characterized by a moderate overcast, and having the same amount of hourly insolation, might be totally different. It might turn on and run for half an hour in the first case, and not at all in the second. Such uncertainty is irreducible, but leads us to view somewhat skeptically the sometimes ascerbic discussions about the exact correlation between direct and diffuse components.

It is much more important to notice that as the sun gets lower in the sky, more and more of the direct beam is attenuated by the increasing air mass, inevitably increasing the importance of the diffuse component. Fig. 3.5 shows this drop-off for clear, sea-level skies.

Another important point to grasp is that the amount of water vapor in the air has an influence not only upon the diffuse radiation, which is increased, but also upon the temperature. Nighttime radiation

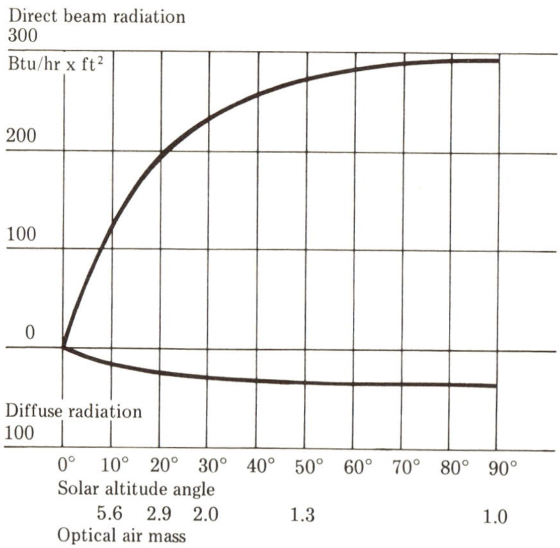

Fig. 3.5 Graph showing the relative drop off of direct beam and diffuse radiation at sea level as the sun sinks lower in the sky. Air mass one occurs with the sun directly overhead. (Data from ASHRAE Publication 170, 1977, page 3-IV.)

cooling is important, both in winter and summer. Humid conditions largely prevent radiation cooling, holding winter and summer night temperatures up.

If sun-pulses are to be useful in predicting the behavior of solar systems, any correlations between ambient temperatures and the distribution of cloudy and sunny days would be of great importance. Surprisingly, there seems to be little. In the author's Sun-Pulse II, complete listings of average ambient temperature, temperature swings, and insolation levels are presented for the full typical year in each of the original 26 TMY weather stations. Examining the area with which he is most familiar (Boston), the correlation between very clear weather and very cold weather in winter, obvious to the most casual New Englander, is not particularly evident in the data. There are cold, clear days; but there are also warm, clear days and cold, cloudy days. Part of this is undoubtedly due to the superimposing of broad-scale weather changes on diurnal swings. In any case, unless such correlations jump out of the data or the designer's experience, one is well advised to ignore them, since the problem of simulating the weather is difficult enough as it is.

As an aside, it is instructive to see the enormous losses in radiation between that which would be available to an orbiting satellite never in eclipse—3,758,000 Btu/year × sq ft—and that available to a tilted collector in Boston—about 450,000 Btu/year × sq ft. It is at least understandable why the orbiting solar power station idea arose.

CLOUDY-DAY SUN-PULSES

If cloudy weather could be modeled with reasonable accuracy by modifying a clear-day sun-pulse, important new avenues would be opened in the design of system simulations. As usual, the hardest problem is deciding what questions to ask. Until we know something about the shape of sun-pulses during cloudy or partly cloudy weather (and this includes foggy, hazy, and overcast days), we really don't know what the problem is. Fig. 3.6 shows a few randomly-selected daily insolation graphs drawn from the 1976 weather data for the Acorn Test House. A little examination shows that several simple things are going on at once to produce a complex output.

1. The day is either clear or turbid, depending upon the condition of the atmosphere, and there may be a layer of continuous cloud cover. These factors define the overall *outline* of the sun-pulse.
2. The day may or may not have intermittent clouds in addition to the basic cover (or lack of it). These will show up as dips in the curve below the general outline.

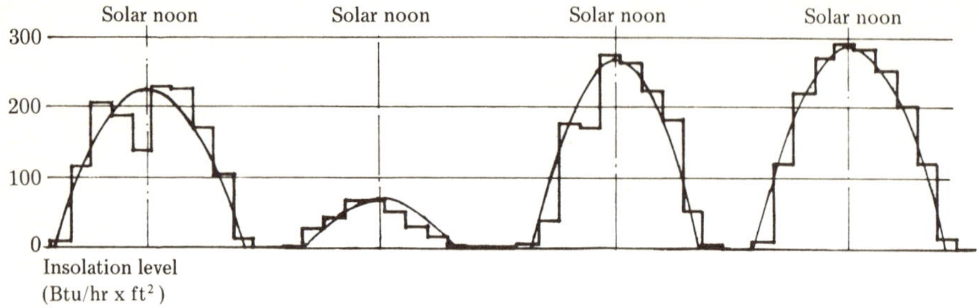

Fig. 3.6 Four successive sun-pulses plotted from Acorn data for January 15th through 18th, 1976 (see Fig. 3.2 for details). The sun-pulse approximations typically are most accurate for very clear and very cloudy days.

3. As weather fronts move in over a period of days, the basic "outline" condition changes, sometimes abruptly with a fast-moving front, more often gradually. The basic outline can therefore change during a day.

Now, if we were dealing with a carefully calibrated piece of machinery with great sensitivity to variations in sunlight, we could well decide to use minute-by-minute data, in order to be sure not to miss the innumerable transient conditions which occur on a regular basis. But solar heated buildings and systems have a strong tendency to average their response to ambient conditions. Since this is so, we need to look ahead and ask which of the many variations in the sunlight actually make a noticeable difference to the system.

First, as we will see, all active collectors have a basic heat loss. Until the collector heats up enough to more than balance that loss, it will not turn on. If the loss rate is a large percentage of the input, we would expect the collector to be off most of the time in cloudy weather, only running to advantage during long, bright, sunny days. This is what happens with inexpensive, relatively high-loss collectors, operating at low temperatures. Exactly the same result will occur with more sophisticated, low-loss collectors operated at high temperatures. One can say, relative to such systems, that only the clear or near-clear days need to be modeled accurately, because the rest of the days produce little or no yield.

On the other hand, low-loss collectors, such as vacuum-tube collectors, have small loss rates at all times. They will turn on during very cloudy weather (one is recorded to have operated during a thunderstorm). On the assumption that they will be on most of the time, the exact shape of the sun-pulse is of little importance, as long as the *duration* and the *total quantity* of impinging energy are accurately modeled.

These are very encouraging reflections, but a third class of system needs more thought, those less sophisticated than vacuum tubes but more sophisticated than simple, high-loss designs. These are likely to be sensitive to the actual amplitude of the pulse during moderately cloudy weather. A day which alternates brighter patches with cloudy ones might or might not result in cycling, depending upon the state of the store. Simplifying the variable amplitude of a real sun-pulse on a partly cloudy day into a smooth sine curve will, for such middle-of-the-road systems, cause some error. The extent of the error is really a function of how many such days occur. It might be noted that, in order to justify solar heating, a substantial number of sunny days are required. In a cloudy climate, one might well select a vacuum-tube collector as the only possible device which can cope with the bad weather.

For most systems, the bright, sunny hours are those during which the majority of the energy will be harvested. Suppose we were to establish a sun-pulse for a bright, clear day, and then reduce the amplitude of the pulse for partly cloudy days until the area under the pulse (and therefore the energy delivered to the collector) is modeled accurately. Would the amplitude of the simpler pulse be correct? The answer can be seen by plotting the average of the two brightest hours during the day against the total insolation for the day, using Boston April 1960 Solmet data, as in Fig. 3.7. In every case for Boston weather, there is a bulge in

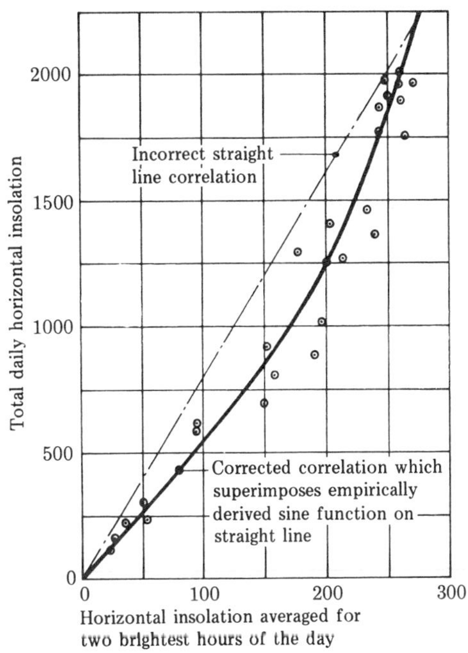

Fig. 3.7 Daily peak insolation plotted against daily total insolation, using Boston April 1960 SOLMET data. The bulge in the curve occurs because partly sunny days are normally not uniformly cloudy. Note that the data clusters on the clearest days, when the collectors harvest the most energy.

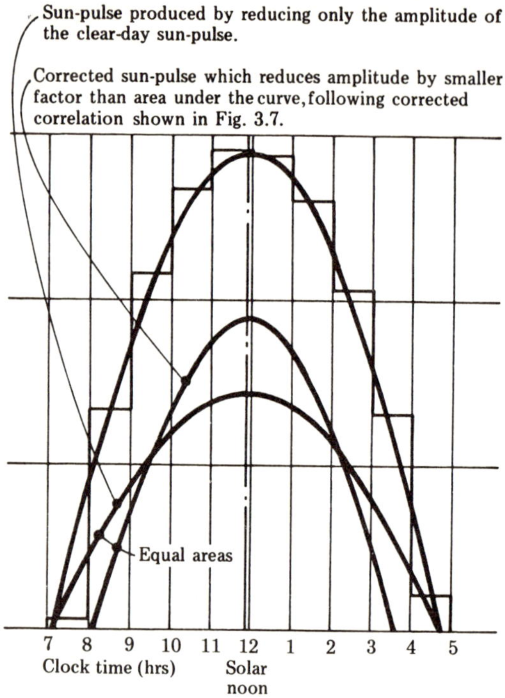

Fig. 3.8 A clear-day sun-pulse and two 50% of clear-day sun-pulses. The steeper cloudy-day sun-pulse falls on the curved line in Fig. 3.7, while the flatter one falls on the straight line. The hourly data from which the clear-day sun-pulse was generated is also shown.

the curve, showing that many of the cloudy days have some hours with almost clear-day insolation, and that even on uniformly overcast days, the cloud cover tends to degrade early and late hours preferentially. The sun-pulse chosen must take account of this (Fig. 3.8).

Suppose the graph showed a day in which the peak two or three hours were equal to clear-day values, while the daily total was half the clear-day value. One would choose to model such days using half a sun-pulse of normal clear-day amplitude, putting together the clear hours in one half and the obscure hours in the other. Such weather pattern seems to be rare in many climates. For instance, no such day occurs in Fig. 3.7. The pattern would be characteristic of foggy periods, periods with a rapid succession of weather fronts, those with frequent afternoon showers, etc.

Days characterized by "popcorn" clouds are beautiful days if you are not thinking about solar collectors. If you are, they can destroy any hope of predicting the output of an active collector. Depending upon the mass of the absorber plate and the built-in dead-band in the controller, the collector might yield nothing (for a massive collector faced with small clouds) or a large fraction of clear-day sun (for a light collector faced with larger, slower clouds). It is reassuring to know that even hourly simulations are helpless with popcorn clouds. Such days will exhibit reduced peak-hour amplitude, along with reduced all-day values; but the actual short peaks between clouds will often be equal to clear-day values.

In the author's Sun-Pulse II Simulation, each place-month of insolation data was modeled by varying the amplitude and area under the average clear-day sun-pulse, taking account of the "bulge" in the data by decreasing the amplitude of the pulse less than the area for partly cloudy days. This results in relatively steeper pulses during such days, which would increase the chances of a collector's turning on. In a simulation using daily data, a more exact imitation of the actual insolation curve could be simulated, with correspondingly more accurate results. Remember, however, that the raw data, collected on an hourly basis, is already strongly averaged and disguises the existence of popcorn clouds. Only by a sophisticated analysis of cloud type and cover can this basic error in the raw data be corrected. It seems to the author that such an analysis is unnecessary, given all the other uncertainties about solar heated systems.

Passive collectors are "on" all the time; thus, the average hourly insolation measures the solar energy received more accurately than it does in the case of active collectors. The details of the rise and fall of insolation levels are still of interest, but for a different reason. Rather than determining the collector's turn-on and turn-off times, changes in the insolation rate influence the rate at which heat flows into storage. Typically, passive systems are less sensitive to insolation level than active ones because of the lack of a turn-on and turn-off control.

THE CORRECTION FOR OFF-ANGLE REFLECTION

In this chapter, we are dealing with the pulse of sunlight *impinging upon* the collector. In the next, we will deal with what happens to this energy *after it hits* the collector. Unfortunately for such a neat separation, there is a step between, which has been included in this chapter even though it relates to the construction of collectors.

Light passing from one medium to another (in this case, from air to the glazing cover of the collector) is to some measure reflected, even if it strikes the surface perpendicularly. The amount of such reflection depends upon the refractive index of the material and the number of covers and will be taken account of in the next chapter. If, however, the light strikes the surface at an angle, additional reflection occurs. This also depends upon the refractive index of the material and the number of covers, but can be approximated by a simple formula which corrects the value for perpendicular sunlight. Standard collector tests now provide the value of the key parameter in the formula, allowing the correction to be made mathematically for any sun angle.

What we need is a way of modifying the basic sun-pulse to account for the off-angle effects. The correction cannot be done exactly by a simple formula, since the reflection depends upon the angle of incidence, which in turn is a complex function of the angle of tilt and the elevation and direction of the sun's direct beam. Suppose, for instance, that the off-angle correction were applied to the sun-pulse for a collector tilted at the angle of the latitude from horizontal. If applied to another collector tilted at a different angle, the angles of incidence would all be different, requiring yet another set of adjustments.

To a good approximation, the angle of incidence is equal to the *hour angle*; that is, the number of degrees the direct beam of the sun is away from perpendicular relative to a tilted collector is about equal to the number of hours the sun is away from true noon, times 15° (360°/24 hours = 15°/hour). The standard off-angle formula, which is accurate for usual angles of incidence, is:

$$\text{Percentage of normal value} = 1 - m\left(\frac{1}{\cos \Theta} - 1\right)$$

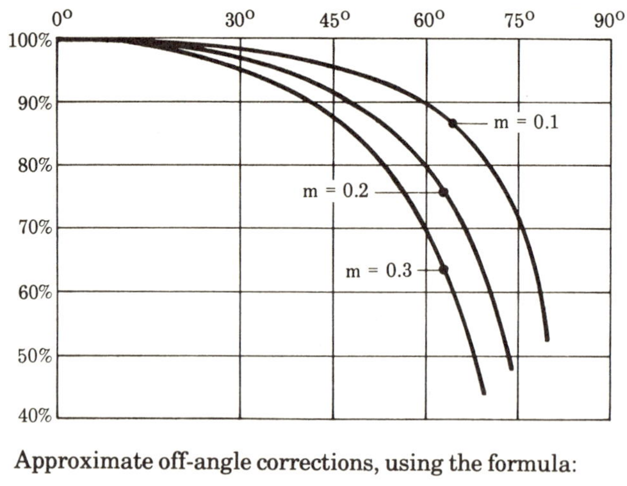

Fig. 3.9 Approximate off-angle corrections, using the formula:

Fraction of radiation at normal incidence $= 1 - m(1/\cos \Theta - 1)$.

Where Θ is the incident angle and m is an empirical coefficient. A single-glazed collector will normally have an $m = 0.1$.

where m varies from about 0.1 for single glazed collectors to about 0.2 for double glazed collectors (see Fig. 3.9). If this equation is solved for the cosine of the hour angle taken at the point where the collector is likely to turn on or off (at an insolation rate of about 80 Btu/hour/ft^2), a new value can be found for the sun-pulse amplitude at that point. The effect is to "squeeze" the sun-pulse a little, reducing its area to account for the off-angle correction losses.

NON-SOUTH SUN-PULSES

If the collector is oriented off-south, the sun-pulse is slightly distorted, and the peak no longer occurs at solar noon. Despite this asymmetry, it is useful to represent an off-south sun-pulse by an approximate true sine wave, as shown in the diagram. The relationship between the peak amplitude and area of a true-south sun-pulse and one oriented at various degrees off-south is shown in the graphs in Fig. 3.10.

The variation in the time of peak insolation causes the relationship between insolation rate, ambient temperature and the heat generated by the occupants to change. For instance, in a collector facing southeast, peak insolation might occur at 10:30 A.M. solar time. Collection then occurs during a cooler part of the day than for a true-south or southwest oriented collector; and the amount of energy usable directly in the building will change due to the change in the time the occupants provide useful heat as an offshoot of their normal activity. However, shifts between occupant time and solar time can be as large as two hours during daylight saving time. Unless these variations are accounted for as well, it is probably unnecessary to worry about the asymmetry of the sun-pulse. Solar heated systems tend to have a lot of inertia, and the overall system behavior is smoothed and averaged, which makes exact modeling of time shifts relatively unimportant.

By far the most important effect of off-south orientation is the reduction in useful gain. In addition, the gain which does occur requires solar access at a lower profile angle than in a true-south orientation; conversely, it is likely that the gain will be further reduced by shading which would not affect a true-south collector.

SHADED SUN-PULSES

It is difficult or impossible to determine with much accuracy the extent of shading from environmental sources. Buildings are erected on adjacent sites; trees leaf out and then become bare-branched; snow blocks collectors on occasion; vegetation grows and dies back. The uncertainty is not in the way reality is modeled, but in the reality itself. Only rough

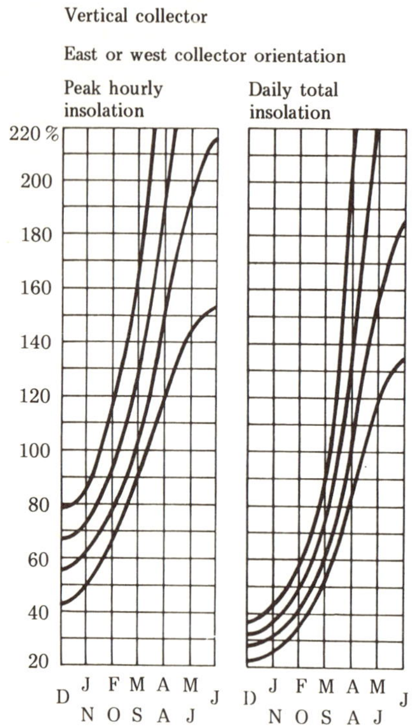

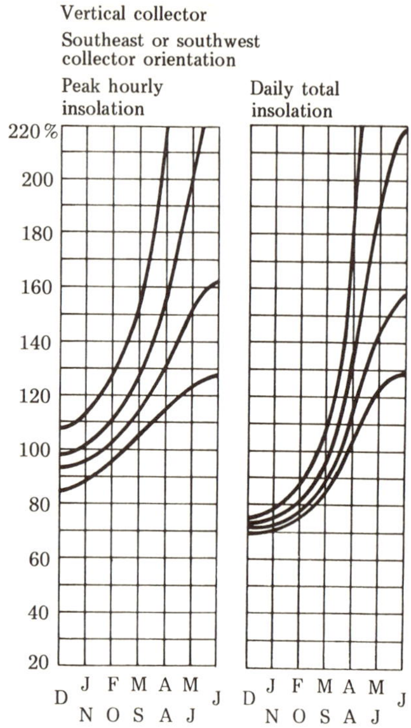

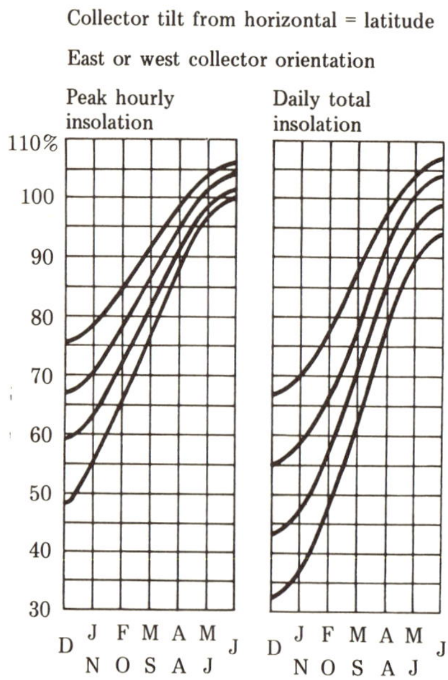

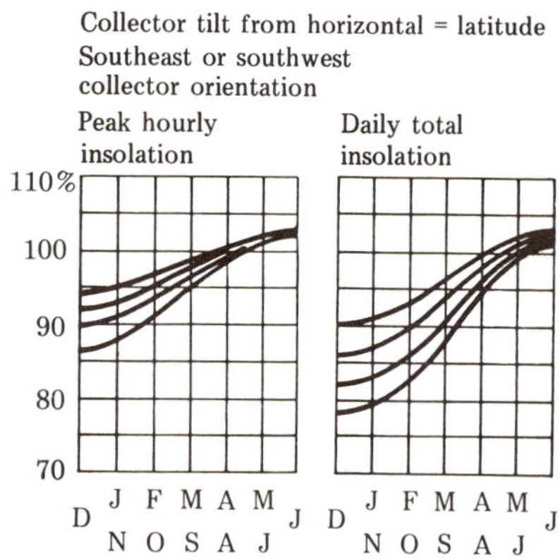

Curves are for latitudes 48°, 40°, 32° and 24° (top to bottom)

Fig. 3.10 Graphs showing the percentage variations in all-day and daily peak insolation for collectors facing 45° and 90° off south, relative to collectors facing true south. Data for two tilts (latitudinal and vertical) and four latitudes is shown. Since in many cases the exact time of day at which peak insolation occurs is unimportant, these curves can be used to modify normal sun-pulse data for collectors at sub-optimal orientations. The data is taken directly from Farber and Morrison's tables in ASHRAE Publication 170, 1977. Since only the relative value of the insolation data is used, the exact values can be somewhat in error without affecting the overall results. Slight variations due to the change in earth-sun distance have been ignored; the data was calculated for the months December through June, and extended to the months July through November.

guesses can be of any use in this sometimes crucial matter. Shading from roof-top structures, or self-shading by adjacent banks of collectors *can* be calculated exactly.

In modeling shading, it is incorrect to adjust the area of the collector, since the heat loss from the entire area of the collector will occur whenever the system is running. Instead, the insolation rate must be adjusted. One advantage of using sun-pulses is that both the peak amplitude and the area under the pulse can be adjusted to account for various kinds of shading. In performing such adjustments, it is useful to realize what happens when one or both of these variables are changed. If the pulse amplitude is changed by the same percentage as the area under the pulse, the wavelength of the pulse stays constant. This would, for instance, model shading which peaked around noon and fell off in the morning and evening. For morning and evening shading only, the amplitude is left alone, and the area changed. This in effect shortens the wavelength or time of day during which sun is available. Reducing the area under the pulse by a greater amount than the amplitude models all-day shading, such as occurs with adjacent banks of collectors. Since the sun-pulse must be adjusted monthly in any case, the shading adjustment can also be adjusted monthly.

It is always worthwhile to attempt to model shading, since it may have an impact not only on the system cost effectiveness, but on the sizing of key components. And, at some point, too much shading will stimulate a responsible designer to abandon solar collection on that particular project.

TILT AND GROUND REFLECTANCE

In calculating a sun-pulse, it is necessary to start with pyranometer data, which will be specific to a particular tilt. Weather service data is available only in horizontal form, since the data was originally accumulated for agricultural purposes. For each tilt of collector, the sun-pulse will be different both in amplitude and in the area under the pulse. While it is possible to compute the exact value of each hour for each tilt, this procedure is time consuming, and is reserved for large main-frame computer simulations. It turns out that if one set of tilted data is generated, it is possible with a small error to translate to other tilts by making the correction proportional to the change in the noontime value, assuming the entire noontime value is direct radiation. The procedure used in Sun-Pulse II was to generate values for tilt = latitude, then add and subtract 15°. In addition, a translation was made to a vertical tilt. For low sun angles, this translation was reasonably accurate; for high angles, it produced very large errors, but in small values. As one moves south, the amount of sun striking a vertical surface becomes quite small, especially in the summer, until, at the Tropic of Cancer, no sun at all strikes a south-facing vertical surface on June 21st. The large error introduced by making a simplified calculation from tilted data occurs only when no useful solar gain would occur, and when the effects of mullion shading and other architectural features becomes large. A more exact translation to vertical data could easily be made.

Ground reflectance is, like shading, difficult to calculate. For those fortunate enough to have a large open field to the south of their vertical collector in a northern climate, very large reflectances can occur. As measured at the Acorn Test House, which had a 47° tilted collector facing a flat open site, there was no detectable ground reflectance effect, even with a bright snow cover. This was due to the very large incident angle of the reflected light. For practical purposes, ground reflectance need only be considered for vertical or near-vertical collectors. Take account of banks of trees, which would cast a shadow on the ground in front of the collector.

Assuming a vertical collector, no shading, and an absolutely flat site, the limiting condition would be an infinite, perfect mirror. The collector would then "see" two identical skies, each producing exactly the same amount of energy: i.e., the radiation would be doubled.

The next extreme condition would be a perfectly white, perfectly diffusing surface, which would produce a new "sky," uniformly illuminated by the amount of radiation from the real sky, producing approximately the same effect as the first case. This upside-down sky radiation is usually characterized by the same formula as uniform diffuse radiation in the real sky, multiplied by a factor representing the percentage by which the actual ground approaches a perfectly white diffusing surface. Numbers like 0.2 are useful for grass, and 0.7 or higher for snow in varying conditions. But remember to discount for shading of various kinds.

Summarizing, if a vertical collector looks out on an infinite plane of snow, one can increase the available insolation by as much as 70%, and if the surface is grass, by 20%. In real situations, trees, shadows and other irregularities cause a large reduction in these ideal values. As the collector is tilted from the vertical, very soon the combination of high incident angles and the reduction in the percentage of the sky dome "seen" in reflection from the ground reduces useful reflectance to an unmeasurable level.

Frequently, schemes are built using specular reflectors placed in front of the collectors. These devices need careful design and calculation, since the plane of the reflector is limited, and the sun bounces off the plane only part of the time. It is important not to encourage sun during periods of excess supply, nor to overheat the collector and cause damage during stagnation. Also, anything specular, unless very well made, will deteriorate over time because of the weather. In snowy climates, the trough inevitably formed between the reflector and a tilted collector will collect snow and inhibit collection. In any climate around trees, debris will also collect in such a trough, causing deterioration. If the reflector is adjustable and manual, someone must remember to adjust it or, if mechanical, the machinery must be maintained. Ice, wind and snow can damage or prevent the operation of large sheets of metal which are movable.

For all these reasons, the author does not recommend reflectors, except for mild climates and in structures where regular maintenance and/or replacement of the reflectors can be tolerated.

A SUMMARY OF THE SUN-PULSE CONCEPT

Solar heating systems including the building which they serve are part of the construction industry, and are therefore subject to all the vagaries and uncertainties inherent in the construction process. From beginning to end of the design and construction of a building, changes are made, misunderstandings occur, materials are not as specified, errors crop up, mistakes are hidden, and innumerable compromises must be worked out to hold down costs. The completed building will be occupied by altogether unpredictable creatures, who will do astonishing things which modify the load, such as use 1/6th as much hot water as you expect (see Chapter 8). The building is bought and sold, maintenance instructions are lost, pieces are added which shade collectors. Finally, like people, the building ages and moves, cracks open up, and materials deteriorate.

Even if perfect information were available about the building and its occupants, a perfect simulation of expected performance is impossible, because the weather will not behave exactly as we expect. We will never have an average year, and the atmosphere is changing constantly as a result of man's industrial processes. Particulates, such as the growing amount of "fugitive dust" created by the endless grinding and rubbing of materials together in industry and transportation, will continue to change the character of radiation at a given site.

Perhaps most important, we have highly-flawed weather data, taken at only a few sites across the country. Even if improved, the data is integrated for hourly or tri-hourly periods, forever eliminating the interesting and sometimes crucial transients which occur on a minute-by-minute basis in the real world. Even if we had such transient data, the response of a collector to rapid changes depends upon its construction, installation and control system, all of which are almost entirely unpredictable on a fine scale. But most crucial of all, weather conditions can change dramatically on a microclimatological level. In the Boston area, it is claimed by some that insolation changes 25% over a ten-mile distance, depending upon whether you live in a river valley or on a hill.

The obvious response to this great uncertainty is to develop design judgment and forget the minute details. To help with this process, the author has developed the idea of summarizing hourly insolation into a series of sine-wave shaped sun-pulses; the amplitude and integrated area under the sun-pulse can be characterized for any month in any location. By further condensing the data into a few characteristic days for each month, and adding the average monthly temperature and the average swing of temperature, an entire year's weather data can be portrayed with sufficient accuracy for most simulation needs in form of twelve numbers for each month.

Furthermore, the data has become so simple that it can be adjusted easily to account for the innumerable special situations encountered at a particular site, such as variations in tilt, orientation, ground reflectance, and shading. Off-angle corrections can be made, and it is easy to adjust the data, based upon crude measurements, to account for local microclimate.

The system works, and can be put into operation today. It would be simple to refine the process of generating and modifying such data, and such refinement can be done on a local basis with local designers who do not have a doctorate in atmospheric physics.

Simple calculation procedures, suitable for small calculators, such as the TI-59 or HP-41C, can make accurate predictions about system behavior using sun-pulses, since the entire sunshine period can be modeled in one time period rather than on an hourly basis. The Sun-Pulse II Simulation is one such procedure.

Finally, as the author hopes will become evident as you read through this book, the sun-pulse idea helps make complex issues easy to visualize. There is some evidence that, for those of us not blessed with a well-developed ability to deal with mathematical abstractions such as differential equations and LaPlace transformations, visual analogies offer the only hope of understanding important characteristics of systems in general, and solar heating systems in particular.

4

THE COLLECTOR

THE OUTPUT EQUATION

In the preceding chapters, we have dissected the incoming solar radiation into its elements:

1. The astronomy of the sun's direct beam, and how it changes through the year
2. The physics of the atmosphere, and the ways in which solar energy is scattered, absorbed, and diffused as it passes through it
3. The fixed relationship between the collector and the sky vault, as measured by its tilt and orientation
4. The procession of clouds and changes in the amount of particulate matter in the atmosphere
5. The idea of the sun-pulse, and its potential for simplifying the innumerable complexities of real weather and solar physics
6. The modification of the clear-day sun-pulse by off-angle corrections, determined by the laws of physics and verified by empirical test

In the modified sun-pulse, the dissected sunlight has been reassembled in a way which permits simple manipulation in a computer program. Our task now is to examine what the collector does with this available energy.

The basis for almost all analyses of active collector response is the simple equation, based on work done nearly 30 years ago by Hottel and Woertz and proved approximately correct through recent tests, both on the stand and in the field. Granted, it is more accurate to use a shallow curve to graph the output of most collectors (see Fig. 4.5); but in order to discuss such subtleties, it is necessary to understand the basic relationships, and this proves easier if the straight-line performance graph is used.

What we are looking for, and what is determined in the standard collector tests (ASHRAE 93-77 is now the accepted industry standard test), is the output of a collector, given the available sunlight entering the cover, and given its temperature and that of its environment. The relation of these variables is shown by the simple equation mentioned above:

Output =

$$\text{Area}\left[\left(\text{sunlight}\right)\left(\begin{matrix}\text{percent transmitted by the cover}\\ \text{plates and absorbed by the absorber}\end{matrix}\right) - \left(\begin{matrix}\text{heat loss coefficient of}\\ \text{the overall collector}\end{matrix}\right)\left(\begin{matrix}\text{temperature}\\ \text{of collector}\end{matrix} - \begin{matrix}\text{temperature}\\ \text{outdoors}\end{matrix}\right)\right]$$

In symbols: $$Q = A[I(\tau\alpha) - U_L(T_c - T_a)]$$

As will be discussed later, the desire for added precision has resulted in the addition of a second order term to the equation, to account for the drop off in output at high temperatures due to radiation. Whether anything is accomplished by this kind of refinement is doubtful, and we will ignore it for the present; we are after principles, not certainty.

It is convenient to analyze the two parts of the equation separately. The first part ($\tau\alpha I$) relates to the *solar radiation* and its losses; the second relates to the *loss of heat.*

COVER AND ABSORPTION LOSSES

The available energy is represented by I, for INcident SOlar RAdiation, condensed into the acronym "insolation," a singularly unfortunate pun on insulation, and one likely to persist despite the efforts to substitute "solation." It is represented here by the amplitude of the sun-pulse at any given moment, and all cosine angles, losses through the atmosphere, and off-angle reflections have already been accounted for by the time you get down to the "I" value. In reading the results of the ASHRAE 93-77 tests, it is important to remember that the insolation falling on the *entire area* of the collector is used; earlier tests only measured insolation falling on the glazed cover plates.

Tau (τ) accounts for the losses to the incident visible, UV, and IR radiation as it passes through the glass or plastic cover(s). A few generalizations can be made:

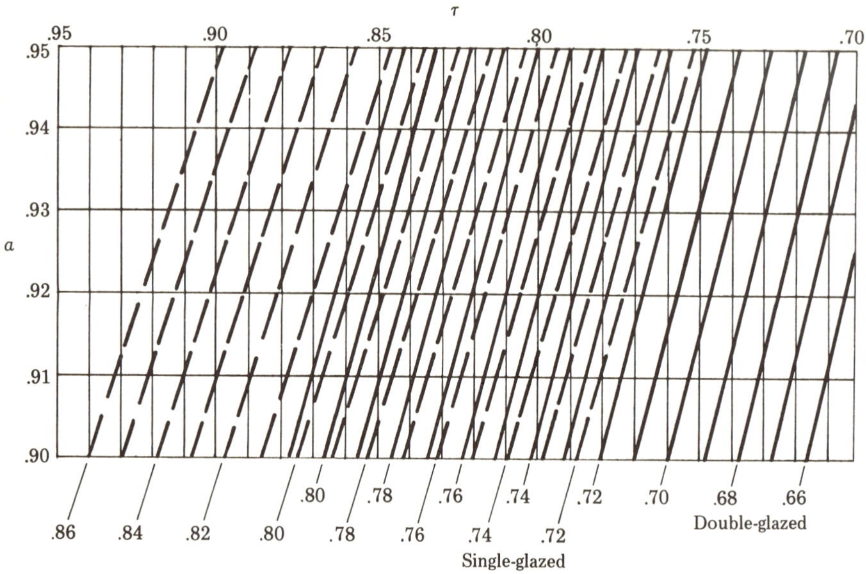

Fig. 4.1 Chart showing the $(\tau\alpha)$ coefficient for common values of τ and α, based upon the formula:

$$(\tau\alpha) = \frac{\tau \times \alpha}{1 - (1 - \alpha)\rho d}$$

Where $\rho d = 0.16$ for single glazing and 0.24 for double glazing. (Chart derived from Duffie and Beckman "Solar Energy Thermal Processes," 1974, Wiley.)

1. Two layers of anything lose a little less than twice the energy of one.
2. Certain impurities in glass, such as iron, absorb parts of the sun's spectrum; removing them increases the value of τ, although such glass is more costly.
3. Thin layers transmit more energy than thick ones.
4. Plastics range over a wide set of τ values. Often, two different plastics work best together for many reasons, one of which is the use of a very thin layer of plastic for the inner glazing, to cut down heat loss, while maximizing the transmission.
5. There is an irreducible reflection loss which is a function of the cover's refraction index. An etched coating on the glass can greatly reduce this loss.

Alpha (α), for absorptivity, accounts for that part of the transmitted insolation absorbed by the plate. Almost all black surfaces have absorptivities of about 95 percent; and a dark room might have an absorptivity of 90 percent. As we will see later, it is the emissivity, or reradiation of heat, which is of primary concern. Relatively unsophisticated paints will give a satisfactory α, providing they will stay on the absorber plate and withstand the necessary stagnation temperature (see later) without giving off gases which will condense on the cover and ruin the transmissivity. Achieving the low emissivities desired to cut down heat loss requires more sophisticated coatings such as black chrome. Again, such coatings must be carefully applied to ensure long life. Because the glazing is to some extent opaque to the heat re-radiated from the absorber, the "combined transmissivity-absorptivity product" or $(\tau\alpha)$ is not exactly equal to $\tau \times \alpha$ (see Fig. 4.1). Often $\tau\alpha$ is intended to mean $(\tau\alpha)$, and the terms are used interchangeably in this book.

At this point, the collector can be considered a heat exchanger; the absorber plate has absorbed a certain percentage of the available insolation and turned it into heat. How does it get to a place where it is useful? How much leaks back outside? These are all questions involving heat transfer.

COLLECTOR HEAT LOSS

There exists in the solar industry a profound dichotomy between those who advocate site-built or passive solar collectors and those who make "hardware": separate, modular collectors which can be shipped and installed onto almost any roof. When you consider collector heat loss, this dichotomy emerges as a key issue, because the collector is in a way an extension of the interior space of the building.

If a collector is built-in, either as a passive collector or as an active one built within the living space (seldom seen, but often thought of), the question of heat loss belongs to the building design. The collector takes its place along with the other windows in the building as a major source of building heat loss. In a modular collector, however, the heat loss is usually considered separately from the building, and accounted for as a reduction in the available solar energy supply. This practice, while it obscures the importance of considering collector heat loss along with all the other system heat losses, is nevertheless logical. Modular collectors, within certain limits, have definable, reasonably constant heat-loss coefficients. Perhaps most important, when the collector is off, it does not lose any useful heat, except for a small amount due to the controller's "dead band" (see later). Since active collectors are off most of the time, it makes sense to isolate their losses from those of the rest of the system, which occur all the time.

As with any part of the building, the collector loses heat to the environment whenever it is warmer than the environment—and, in fact, gains heat from the environment if it is cooler (a rare occasion in glazed collectors, but very common for unglazed swimming pool heaters operating in hot weather). In a modular collector, this loss is *additional* to the loss which occurs in the house itself, whereas in a passive direct-gain design (see Chapter 7), it is argued that the windows would be there *anyway*. If this is the case, it is a legitimate difference. But in many cases, a passive design has substantially more glass than the owner might have been satisfied with otherwise. The glass is expensive, and the solar gain must pay its way. Often, because of overheating in the space and residually high temperatures at night, the passive design will have a greater heat loss (especially if overheating is solved by dumping warm air and introducing cooler outside air) than a corresponding building with an ordinary amount of glass. Because of this, it is sometimes the case that a well-designed active system yields a payback comparable to that of a passive one. Passive designs have compensating virtues, such as architectural character, lack of leakage, and, perhaps most important for the architect, a need for building design skills rather than mechanical engineering. Furthermore, it is much more difficult to cut passive elements out of a building budget, since they perform other architectural functions.

If the output of a collector is taken as a percentage of the impinging radiation, and treated as an efficiency, the collector can exhibit efficiencies less than zero and greater than one. Evidently, the output divided by the input is *not* an efficiency at all, although it is commonly so called. Up to the point at which heat is embedded in the absorber plate, it is legitimate to consider efficiencies, with reflection, shading, dirt, transmission, and absorption losses contributing to a fractional efficiency of solar-thermal conversion. In dealing with the heat-transfer and heat-loss components of collector performance, however, we are dealing with an extension of the building heating apparatus. As we will see, the heat loss from a collector is as much determined by what goes on in the building as by the construction of the collector. Collector heat loss is properly a *system* variable, and will be so treated in this book. This distinction helps to bring together active and passive designs, and if anything is needed in the industry, it is a release from the artificial and destructive competition between the two techniques.

The heat loss of the collector, like that of a house, can be represented by a "U" factor multiplied by the collector area. In collectors, this U-factor is an empirical number, derived by measuring the overall loss from the collector on the test stand and dividing it by the nominal area. In the basic performance equation, the U-factor is in turn multiplied by the nominal area, giving back the heat loss in its entirety. This is an important point to note, since the U-factor can change depending upon how the collector is measured. In the "old" NBS test, the net area of the collector was used, while in the ASHRAE 93-77 test, the gross area is

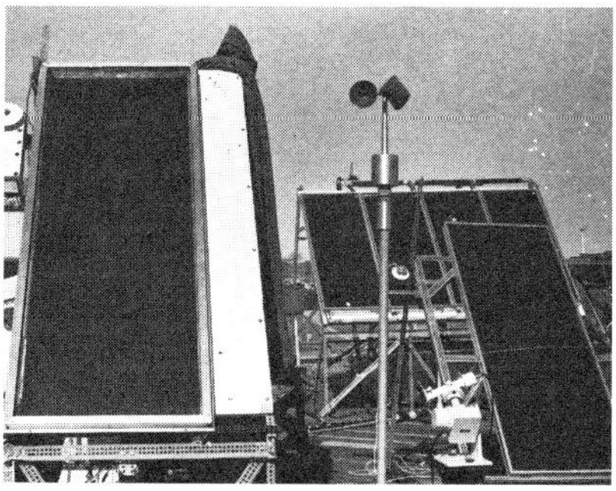

Fig. 4.2 Photograph of collectors on the roof-top testing apparatus operated by the Energy Center at the University of Connecticut, Storrs, CT. Prevailing techniques for testing are designed primarily for relatively small, modular, collector panels. Larger all-roof or site built systems will require a different approach.

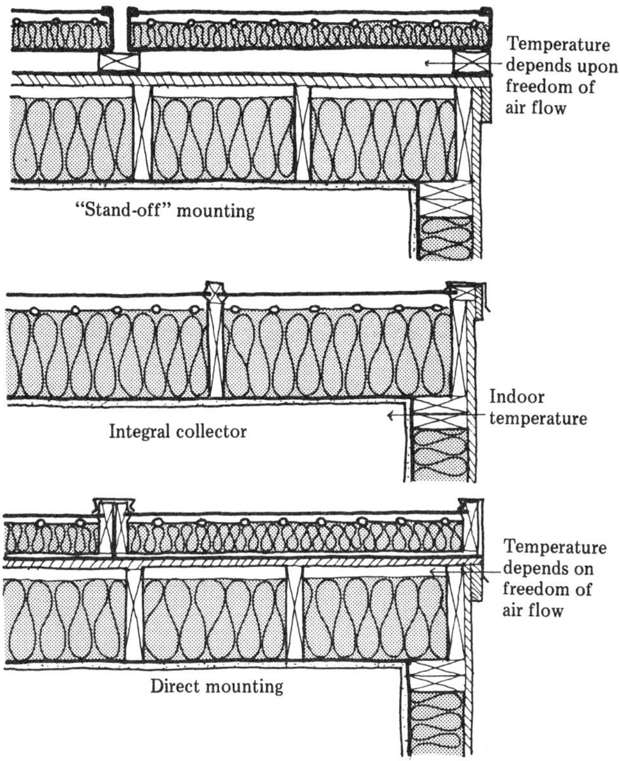

Fig. 4.3 Sections taken looking up the roof, showing various mounting conditions for liquid flat-plate collectors. The back losses are changed substantially by changes in mounting conditions.

used. The heat loss remains the same regardless of how you measure the collector, so the U-factor must change. (It is smaller in the ASHRAE test than in the NBS test.)

By far the largest loss for most active collectors is through the cover. The hot absorber plate emits thermal radiation, which passes through the transparent air above the plate and hits the cover. Depending upon the material of the cover (whether glass or one of the various kinds of plastic), it will totally or partially absorb the radiation, in turn reradiating it in all directions. Much of this radiated heat is lost.

In addition, the air above the plate is heated by conduction from the plate, generating convective motion in the air, transferring heat from the plate to the cover.

The hot absorber plate also loses heat out the back of the collector, depending upon how dry, well-sealed, and well-insulated the back is (see Fig. 4.3). Collectors tested on a stand have the backs exposed to ambient temperatures, representing an approximately accurate model of a collector mounted slightly off a roof, on a "stand-off" mounting (usually required for modular collectors to prevent moisture build-up beneath the collector). If the collector is to be mounted *into* the roof, so that heat escaping from the back "sees" only room temperature air, the heat-loss coefficient of the collector will be slightly lower in operation than in the test. Note, however, that even if mounted onto the roof and flashed in, the collector may lose heat to air which is nearly at ambient. It is common, and good, practice to allow a ventilated air space between the top of the insulation in the roof and the roof sheathing. It would be a fallacy to omit the insulation below the absorber plate if this ventilation air space were left in the construction. While much lower than the cover loss, the back loss is not trivial; it can be serious if the insulation becomes wet for any reason.

Side losses are the least important, but again are not entirely trivial. Most modular collectors have a small amount of side insulation. If a collector is intended for side-by-side mounting, so that adjacent modules lose heat to each other, the test will again overestimate the losses, since the collector on the test stand loses heat on all sides.

Reducing the side and back losses is a relatively straightforward matter of adding insulation and keeping out water and air. Since the insulation for the cover must be transparent, the problem is much more complex and has led investigators to a variety of ingenious solutions.

A head-on approach is to use multiple layers of glazing. Double-glazing, a technique commonly used between 1974 and 1977, is being abandoned in newer models. Tempered glass is expensive, especially if both are low-iron glass. Glass is also heavy; and if sealed double-glazing is used, care must be taken in installing the cover to prevent leakage. But the two-layer idea, using combinations of different glazing, is ideal in many situations. A common design uses an outer layer of fiberglass, which has a relatively low transmission coefficient but is tough and durable; and an inner layer of thin, heat-resistant film, such as Teflon. Glass and Teflon are also used together. Some manufacturers, and lots

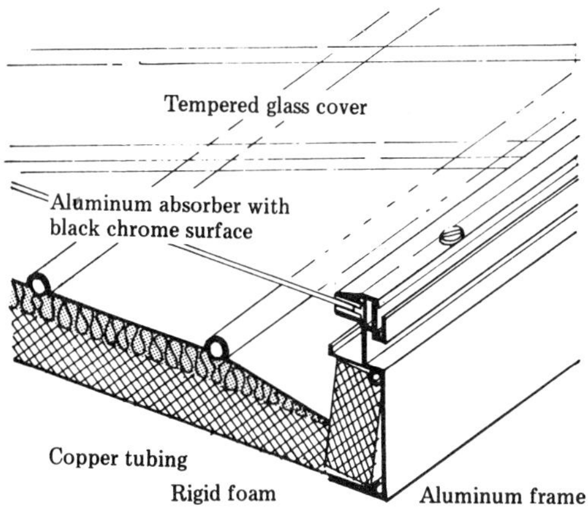

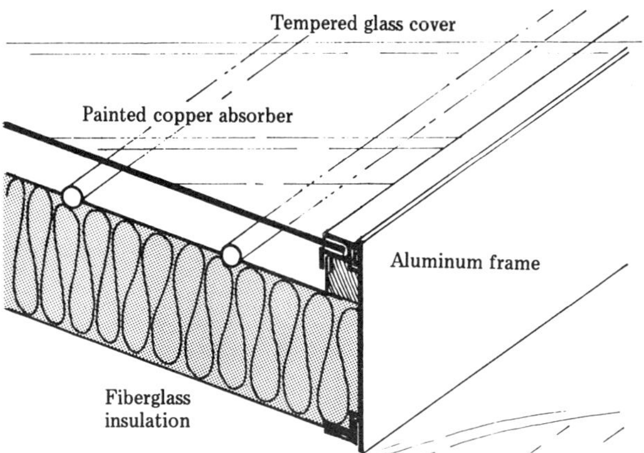

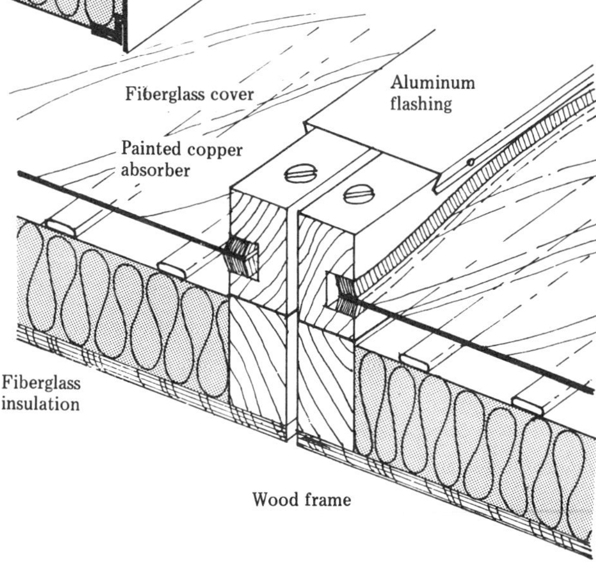

Fig. 4.4 Sections through three representative designs of liquid-cooled flat-plate collectors. Although the potential variety is endless, many brands are very much alike.

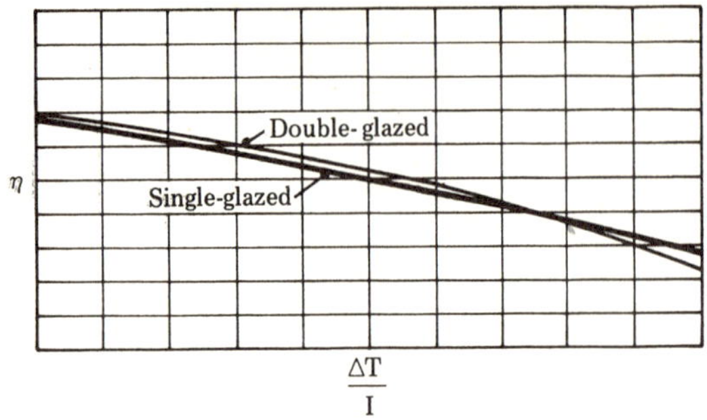

Fig. 4.5 Comparison of the performance curves (based on the ASHRAE 93-77 test) of a single- and a double-glazed collector with black chrome absorbers, all other details being identical.

of owner-builders, are experimenting with two layers of film, such as an outer layer of Tedlar, which resists UV and is tough, and an inner layer of Teflon. Film is highly transparent but is puncturable, hard to install smooth, takes an initial shrinkage upon heating, in many cases is degraded by heat and/or U.V., and "leaks" lots of radiation. This latter characteristic can be a virtue, as we will see later, in reducing the collector's stagnation temperature.

A more subtle approach, but one amenable primarily to absorber plates which can be processed in a factory, is the use of a "selective surface." Volumes have been written on the technical aspects of such surfaces, but their nature can be described briefly. While highly absorbent to radiation at wavelengths in and near the visible region, these surfaces are poor emitters of the longer wavelength radiation arising from a warm absorber plate. Black chrome is one of the best of these, if properly applied, having an absorptivity of 95 percent and an emissivity of about 10 percent. The effect of a good selective surface is almost precisely that of adding another layer of glass. Once having added the selective surface, which is generally cheaper than a layer of glass, adding the second layer of glass is a waste of time, as shown in Fig. 4.5 which compares double- and single-glazed collectors, both with selective coatings.

An even more sophisticated measure requires a total redesign of the collector itself: placing the absorber plate within a thermos bottle. Several companies have developed vacuum tube collectors, using both liquid and air as a medium (see Fig. 4.6). In general, the vacuum tube is a superb heat-loss inhibitor. When used in conjunction with a selective surface, a heat loss coefficient much lower than that of an ordinary single-glazed black-painted collector can be achieved. Because they use standard fluorescent tubing, vacuum tube devices hold promise of being mass-produced for a low price. They remain to be proved in a residential market, or even a low-maintenance commercial market, for several reasons:

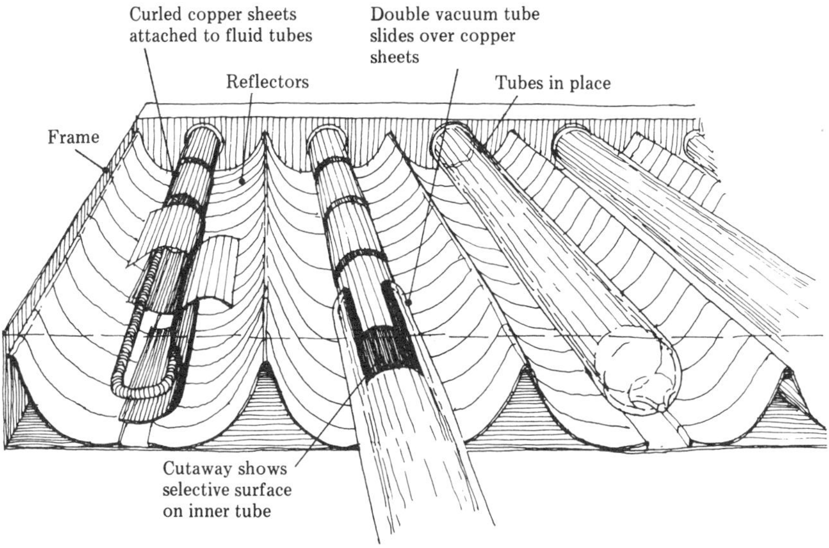

Fig. 4.6 Cutaway perspective of the General Electric "Solartron" TC-100 vacuum tube collector, based upon a 1907 patent.

1. They may need an additional cover to prevent the buildup of debris, birds' nests, and ice between the tubing (which remains virtually at ambient temperature during operation). This cover could, however, be a cheap layer of film.
2. Almost all designs use a reflector behind the tubing, which may suffer environmental degradation.
3. Breakage may be a problem. Designs which do not allow easy removal and replacement of the tubing may be difficult to maintain.
4. Very high temperatures are often used (although there is nothing about the vacuum tube concept which requires this). Such temperatures are a major problem in residences.
5. Since they have very high stagnation temperatures, heat-dump measures are necessary.

Another approach uses vacuum tubes placed side-by-side to form a well-insulated cover plate.

Since this book deals with the general theory of solar heated systems, there is nothing about a vacuum tube device which violates any of the principles developed here. However, because they are so insensitive to heat loss, systems using vacuum tubes can have smaller storage tanks and smaller load heat exchangers than "ordinary" systems. Also, because of the high temperatures possible, a vacuum tube device might very well be able to produce solar cooling at a reasonable cost, and it certainly is one of the few potentially practical routes to solar process heating. Despite the possibility that packaged systems using vacuum tubes and produced by our largest manufacturers may become the industry standard in the 1980's, many people will still use "low-tech" devices, either because they do the work themselves, or because the latter remain more cost-effective. The major result of the successful development of vacuum tube collectors might be the elimination of the collectors in the middle, polarizing the industry still further into high- and low-tech options.

Because they are designed to produce high temperatures, tracking and movable concentrating collectors are also characterized by very low heat-loss coefficients (Fig. 4.8). Typically, such devices require substantial maintenance or complex mounting and are likely to be used primarily in process heating or cooling applications which take advantage of their capacity to produce high temperatures. It is unlikely that movable collectors capable of producing high temperatures will be used in applications requiring low temperatures except where, like vacuum tube designs, these collectors can solve a tight space problem by reducing store and collector size. Concentrators "see" a different sun-pulse from flat-plate collectors, since they can typically use only the direct beam component of the sunlight. For this reason, they are not considered in this book.

Fixed concentrators, which are like flat-plate collectors because they do not track the sun and because they pick up some diffuse radiation, usually require periodic adjustment. For this reason, their use in fixed collectors will be limited to those with wide acceptance angles and low concentrations. They are being used as booster reflectors behind vacuum tube collectors which have spaces between the tubes.

COLLECTOR HEAT TRANSFER COEFFICIENT

To measure the heat loss of a collector, you must know both the temperature of the collector plate and ambient (outside air) temperature. If some parts of the plate are warmer than others, the overall heat loss will be determined by the heat loss coefficient, U_L, and by the *average* temperature of the plate. But we are interested in relating heat loss to the temperature of the fluid entering the collector, which is going to be cooler than the average plate temperature. The heat transfer coefficient, F_R, allows us to relate heat loss to incoming fluid temperature.

Beside its primary function of converting visible radiation into thermal radiation, the collector acts as a heat exchanger, with only one

Fig. 4.7 Cutaway perspective of a sunspace using the patented "Star Light"™ system, produced by the McClintock Energy Corporation of Boston, MA. A double-layer sandwich of plastic is sealed between two layers of glass, with the spaces between the plastic sheets manifolded so that fluid can be pumped through the glazing sandwich. Direct passive solar gain is admitted to the sunspace, while colored fluid is pumped through the glazing from storage to provide active space and water heating.

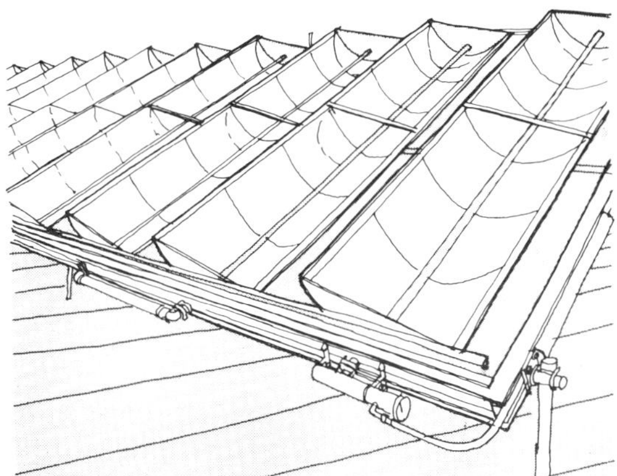

Fig. 4.8 A sketch of a typical rack-mounted movable concentrating collector, wih parabolic trough reflectors rotating on a north-south axis under the control of a heliostat or microprocessor. The troughs rotate around the fluid tubes, which are stationary.

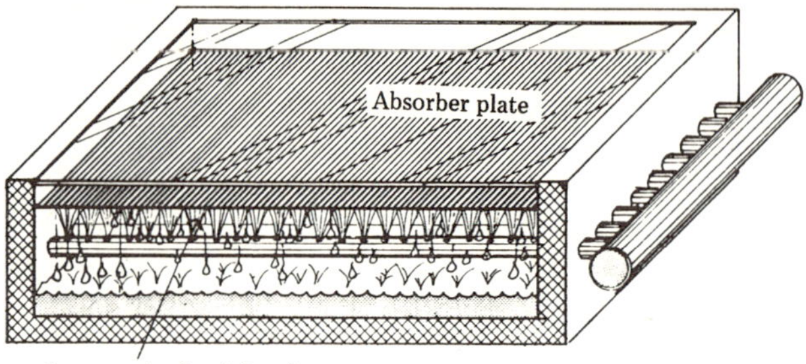

Fig. 4.9 A visualization of the imaginary "ideal" collector discussed in the text.

fluid. (It is important to note that a number of designs use a dark liquid, in which radiation is absorbed directly by the fluid. The behavior of such a collector may differ from that of an "ordinary" collector with a black absorber plate.) In order to simplify the problem, we need to dissect it. We will deal first with the basic heat transfer behavior of the plate, then with the distribution of heat across the plate.

Assume for a moment that the collector fluid is sprayed evenly over the back of the plate (as in some experimental designs). Neglecting the effect of stratification in the air space above the plate from the top to the bottom of the collector, we can assume that the entire plate is at the same temperature while operating. In the limiting "ideal" case, an infinite amount of fluid is sprayed at the back, withdrawing all the available heat. The temperature of the fluid coming in, that going out, and the plate are all equal. The plate, since this is a magic device, is infinitely thin. The first thing to notice is that the collector *still* loses heat to the environment, and that no strategy concerning the *heat transfer* properties of the plate can reduce losses further than the infinitely thin plate with an infinite amount of fluid spread evenly over the plate. The heat loss in this first "ideal" design is equal to the collector U_L times the difference between operating (= incoming = outgoing) temperature and ambient (see Fig. 4.9).

Now maintain the spray-on-the-back design, and substitute real fluid flow rates but neglect the plate thickness. As in any real heat exchanger, the fluid must gain heat in order for the plate to lose heat. The fluid must stay on the plate for a period of time before dripping off, and during that time it must get warmer at the expense of the plate. What is the temperature of the plate? The very thin plate is at the same temperature as the water, and the water begins at incoming temperature and reaches, after a time, outgoing temperature. The average temperature of the plate is therefore the *time average* of the temperature of the water. It will not be exactly the mean between incoming and

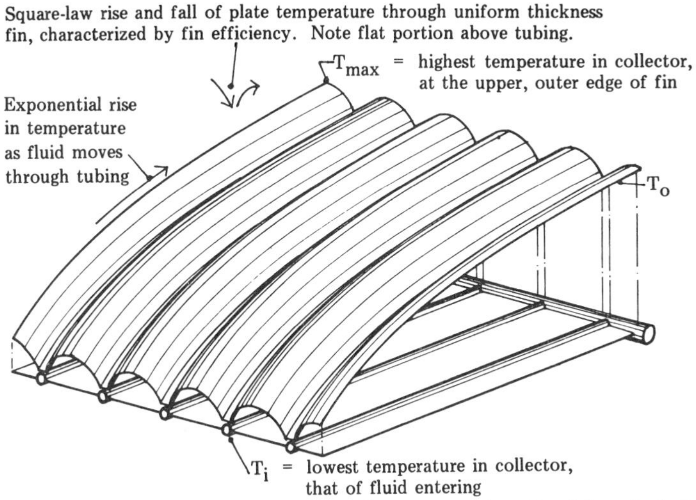

Fig. 4.10 Temperature surface map of an operating liquid-cooled flat-plate collector. The average temperature of the collector is the height of the surface integrated over the entire plane. The concept was developed by Dr. William Beckman of the University of Wisconsin.

outgoing fluid, because the temperature rise will be an exponential curve; but for all practical purposes, one can use the mean to describe the plate temperature.

Now look at a plate with a real thickness. The underside of the plate will be at the average water temperature, plus some amount determined by the power needed to force energy through the insulating film of still water at the interface between plate and fluid (just as the outer layer of still air insulates the wall of a house). Then, the top surface of the plate, where most of the losses originate, will be at a still higher temperature, to provide the power needed to drive energy through the thermal resistance of the plate. While the plate is a very good conductor, nonetheless there will be a temperature drop through it, as there is through all real materials at ordinary temperatures.

Look now at a standard tube and plate collector (see "temperature surface" diagram in Fig. 4.10). Heat absorbed by the outer part of the fin must move horizontally to the tube, through the fin. The rate depends on:

- fin thickness and thermal conductivity (usually considered together as the "*kt*" product)
- distance between tubes

A rough formula for the "fin-efficiency" of a typical tube and plate

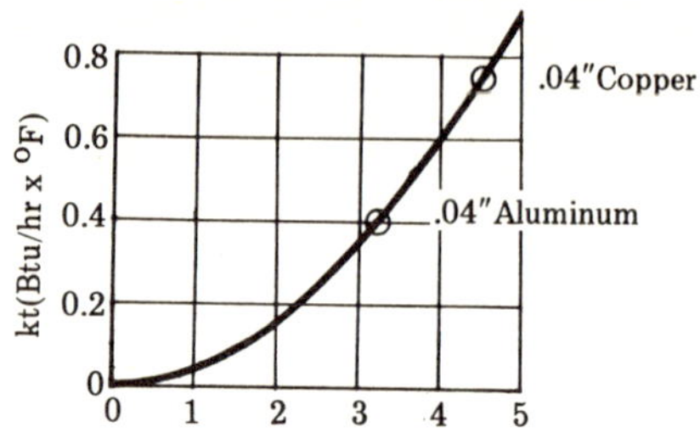

Fig. 4.11 Distance required between fluid tubes for a uniform thickness fin to achieve a fin efficiency of 0.98.

collector is:

$$\text{Fin efficiency } 1.1 - .27\left(\frac{\text{width}}{2}\sqrt{U_L/kt}\right)$$

$$.95 > \text{valid} > .75$$

Note that the overall collector heat loss figures into the formula. This is due to the fact that, if heat is impeded from flowing through the fin, the fin will heat up, increasing the heat loss (remember, we are trying to relate the average temperature of the operating collector to the temperature of the incoming water). However, the fin efficiency is a weak function of U_L and a strong function of kt and the width of the fin. Fig. 4.11 shows the balance needed between width and kt to yield a good fin efficiency.

Before heat reaches the water stream, it must bridge any gap between the fin and tube. While the literature establishes the importance of making a good bond between fin and tube, there is room here for cost-effectiveness analysis. Granted that tubes wired or clipped to plates create a huge resistance to heat flow, clamped-on fins (if secured to prevent the fin from coming off in shipping and installation) give adequate heat transfer at a very low cost. However, the recent availability of well-bonded low-cost tube and sheet combinations are likely to eliminate press-fit designs. The design of tube to plate joints has occupied collector designers for years, and is the subject of a vast literature.

The heat must also flow through the tube (usually a negligible loss) and pass into the fluid stream. The difference in transfer found between laminar flow and turbulent forced flow is quite large and has a strong effect on the temperature difference between the inside surface of the tube and the water (which is what we are after). In most liquid-cooled collectors, the flow is laminar even at relatively high flow rates. Increased flow rate *does* affect heat transfer in the direction of fluid flow, as explained below.

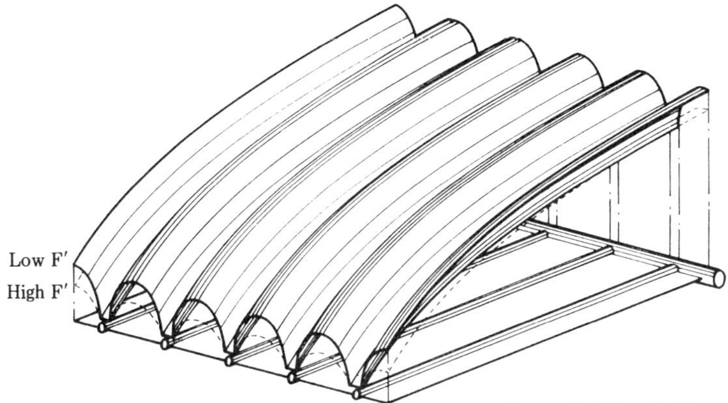

Fig. 4.12 Change in the temperature surface map shown in Fig. 4.10 as F', the collector efficiency factor, is reduced. As F' approaches unity, the surface *across* the tubes becomes flat.

The influence of all the impediments to heat flow in the direction *across* the collector is usually merged into a single parameter called F', the "collector efficiency factor." As the value of this parameter increases, the temperature surface changes as shown in Fig. 4.12.

Once the absorber plate construction is set, F' is *basically* set. But increases or decreases in the collector heat loss *do* influence this parameter, because F' is really the ratio of how well heat moves horizontally relative to losses from the top and bottom of the absorber. As these losses increase, the ratio must change. (F' gets smaller as U_L gets larger.)

Another factor, sometimes called F'', denotes the influence of the rate at which heat is removed by the fluid stream, after it has passed through the boundary layer of the fluid next to the tube wall. Looking again at the temperature surface diagram, the fluid gains temperature *exponentially* as it flows through the tube. (Exponentials are explained more fully in the chapter on Heat Exchangers.) If more fluid, or a fluid with higher capacitance (water instead of antifreeze, liquid instead of air) is passed through the tubes, the effect is to flatten the curve, reducing the height of the temperature surface, as shown in Fig. 4.13. Here, as opposed to parameter F', the *system* design has some influence. Providing the *collector* designer has made the manifold piping large enough, the *system* designer can force more fluid through the tubes and add a few percent to the output. Economic trade-offs comparing pump energy to the extra energy harvest show that high flow rates are quite cost effective. The use of water instead of antifreeze also will increase the harvest.

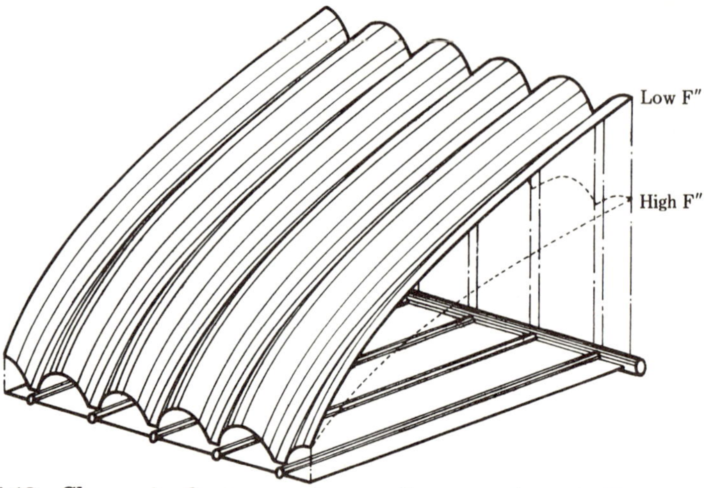

Fig. 4.13 Change in the temperature surface map shown in Fig. 4.10 as F'', determined by the heat capacity rate of the fluid (for a given collector) decreases. The surface will flatten in the direction of flow as F'' approaches unity.

Beside the capacitance and rate of flow of the fluid, the length of the trip through the collector influences the temperature rise—the longer the trip, the greater the rise. But, short collectors imply lots of manifold piping. A trade-off can be made between flow rate and collector length.

Putting F' and F'' together yields F_R, the overall collector "heat removal factor." Fig. 4.14 shows how F_R changes for a given collector design for various flow rates. Notice that F_R approaches F' asymptotically—there is nothing one can do with the F'' parameter which can reduce F'. As F'' approaches 100 percent, F_R approaches F'.

Summarizing, there are a number of ways to increase F_R. First, the absorber plate can be designed to minimize the temperature differences from one point to another. The fluid passages can be made closer, the plate thicker (up to a point), the material better (copper rather than aluminum, say), the joint between passage and plate tighter (perhaps the two are integral). Next, provision can be made for a larger flow of fluid. Notice that the collector designer can only *make provision for* a large flow. The system designer is responsible for insuring that the *actual* flow is adequate. One way to allow larger flows is to make the manifold pipe large enough—1″ diameter rather than 3/4″ makes a large difference if several collectors are manifolded together. The length of travel can be decreased, which allows more fluid to flow per square foot of collector. The final measure is to design the collector to use water rather than antifreeze, at the option of the designer. Water always has a higher capacity to transfer heat than antifreeze. Using it can often increase the F_R by 2 or 3 percent, and this is a gain well worth having.

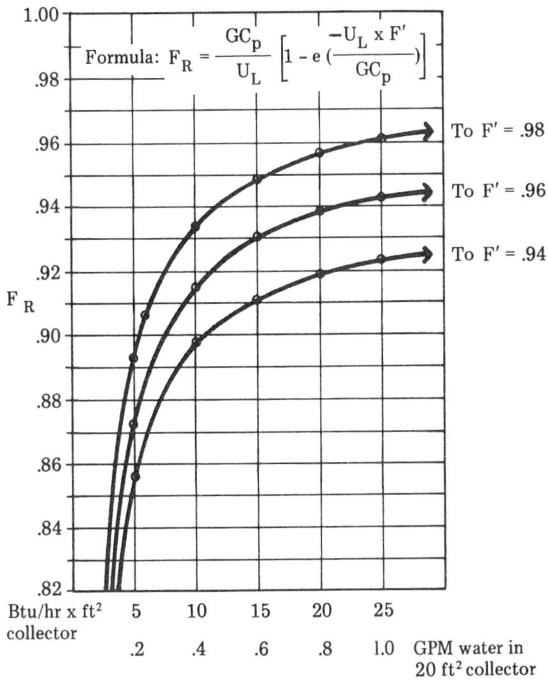

Fig. 4.14 Graph showing the effect of changes in the flow rate and capacitance of fluid on the parameter F_R, for various values of F'. Note that the curves are asymptotic to F' at infinite GC_p values. (From the graph on page 7.7.1, Duffie and Beckman, "Solar Energy Thermal Processes," 1974.)

As a practical consequence, it is often the case that a cheaper absorber plate design, with less elaborate precautions taken about its F', can be compensated for by raising the flow rate, with an *overall* increase in output relative to a better but substantially more expensive collector. Careful cost optimizing is likely to make the difference between really cheap and widely used solar heating systems, and carefully made but excessively expensive ones.

Much of the above discussion is relevant to other collector geometries. Where fluid is exposed to the space between the cover and the absorber, either in some air collectors or in trickle-down collectors, losses through infiltration, condensation, and evaporation may occur, degrading collector performance.

For air collectors in general, the heat-transfer coefficient between absorber and fluid is substantially lower than that in liquid collectors. This effect is partially compensated for by enlarging the area of contact.

Air has a specific heat of about 0.2, so that the heat capacity rate (mass flow rate times capacitance) of an air collector is generally low relative to a liquid collector. While values of 25 Btu/hr × sf(1/20 GPM/sf collector) should be considered in liquid collectors, values above 2.2 Btu/hr × sf(2 CFM/sf) are seldom practical in air collectors. The combination of a very low heat capacity rate and poor heat transfer between the plate and air stream results in F_R's typically as low as 0.6 to 0.7.

GRAPHING COLLECTOR PERFORMANCE

Off-angle reflection, transmission losses, absorption losses, heat losses, and the heat-removal factor—each piece of the puzzle has been explained, and we are now ready to examine the overall behavior of a flat plate collector, as one after the other of the crucial variables are changed in value.

First, we rewrite the basic output equation to account for F_R:

$$Q = AF_R(\tau\alpha I - U_L(T_i - T_a))$$

Solar engineers, used to the manipulation of equations, factor the equation to come up with a convenient graph of collector output (dropping the A, to deal with a unit area):

$$Q = F_R\tau\alpha I - F_R U_L(T_i - T_a)$$

$$\frac{Q}{I} = \frac{F_R\tau\alpha \not{I}}{\not{I}} - \frac{F_R U_L(T_i - T_a)}{I}$$

By giving the key elements names, everyone can talk about the same thing in shorthand.

$$F_R\tau\alpha = e_0 \qquad (T_i - T_a) = \Delta T \qquad \frac{Q}{I} = \eta$$

and we end up with:

$$\eta = e_0 - F_R U_L \frac{\Delta T}{I}$$

The key variables become the e_0, or initial efficiency; $F_R U_L$, the heat-loss coefficient modified by the F_R coefficient; and the $\Delta T/I$. A typical graph looks like that shown in Fig. 4.15.

While meaningful and useful to the engineer, all these aggregations of variables take us very far away from the understandable components discussed in earlier sections. In order to make sense out of the equation, we will use simpler graphs.

The first such graph is a plot of the output of a collector in British thermal units per hour (Btu/hr), as a function of the difference between the actual collector temperature and outdoor (ambient) temperature. In order to do this, we will have to fix the insolation rate, and for this example let us assume it is 250 Btu/hr × sf. (We are dealing with one

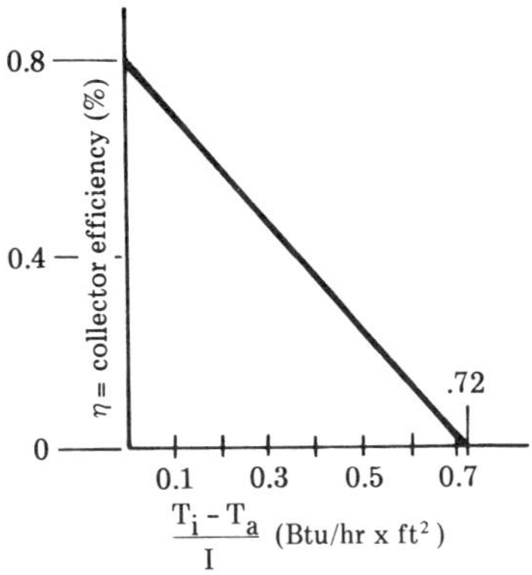

Fig. 4.15 Typical linear performance curve for a flat-plate collector.

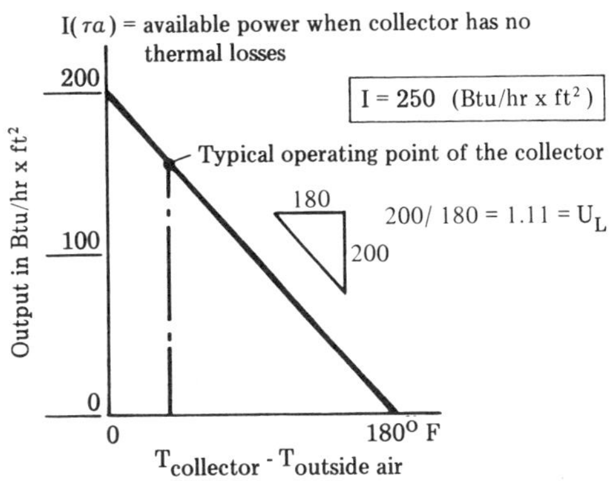

Fig. 4.16 Performance curve for typical flat-plate collector, drawn for a fixed level of insolation, and using the average collector temperature.

square foot of collector.) Let us assume that the $\tau\alpha$ product is equal to 0.8, and that U_L is equal to 1.11. (These values are not far from real values for an actual collector.) The resulting graph is shown in Fig. 4.16.

The heat-loss coefficient is reflected in the graph by the slope of the performance line. This is seen by looking at the extremes (always a good way to understand a graph). If the ΔT is equal to zero, there is no heat loss, and the output is equal to e_0. On the other hand, if the ΔT is equal to 180°F, all of the heat collected has been dissipated to the environment, and none has been taken back to storage. The output is zero, and the collector is in what is called "stagnation." This is a condition which actually exists in the collector when no fluid is flowing through it.

A very important assumption has been made to allow the drafting of graphs such as these, namely that the collector U_L is constant. But this is far from the case, as shown in Fig. 4.17. In this example of a single-glazed collector, wind makes a very pronounced difference; an 11-mph-wind reduces the output on this very sunny day by 8 percent at the low ΔT of 54°F, and by 21 percent at a 90°F ΔT. Conversely, if the

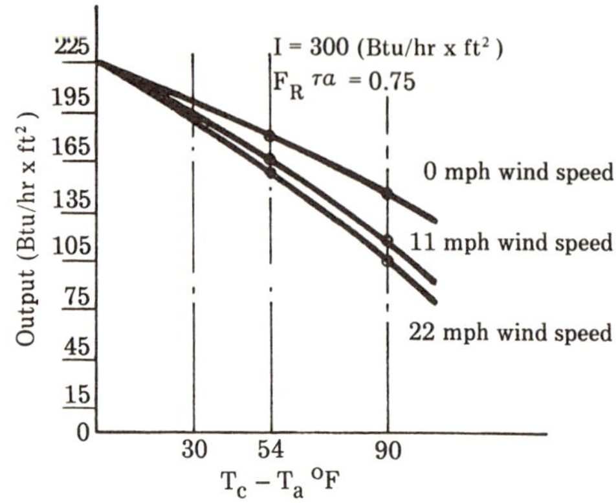

Fig. 4.17 Graph showing the theoretical variation in output for a single-glazed collector, assuming all losses are through the cover, for various wind speeds. The wind strips off the outer air film and degrades the collector U_L.

graph for a windy day is used, performance will increase dramatically on still days.

In addition, the graph curves slightly, due to the nonlinearity of the radiation component of the loss, increasing the U factor by 20 percent or so between 0° ΔT and 90° ΔT. The effect is much more pronounced in plastic-glazed collectors, which are partially transparent to infrared radiation.

Finally, not shown on the graph, the U_L factor rises slightly with the plate temperature, regardless of the ΔT (a negligible effect).

In writing standards to determine collector output, the industry at first decided to establish a fixed, linear performance curve for collectors. Now, both a linear and a nonlinear approximation are required. Manufacturers naturally try to get their collectors tested at the lowest possible wind speed, leaving the consumer to wonder what performance he can expect. But we do not need added complexity: wind can be factored in later, if the basis for the performance test is known. And the nonlinearity can usually be ignored for well-designed systems operating at low temperatures. The message is clear: heat loss data is not known to a high level of precision.

Let us now accept the linear, constant graph of U_L and play with the variables. First, we must convert the graph to use the temperature of the inlet water (T_i) rather than that of the collector as a whole (T_c), since we know the former and not the latter.

F_R does the trick, and its effect can be shown by introducing a *tilted* set of lines in Fig. 4.18. In our ideal collector, the two sets would be the same, since $T_i = T_c$. The vertical set can therefore be thought of as defining the loss of an ideal collector, with the tilted lines adding in the loss due to the real drop in temperature through the absorber and fluid stream. The ratio of the output *after* this loss to that before is F_R. The higher the F_R, the more tilted the $T_i - T_a$ lines are.

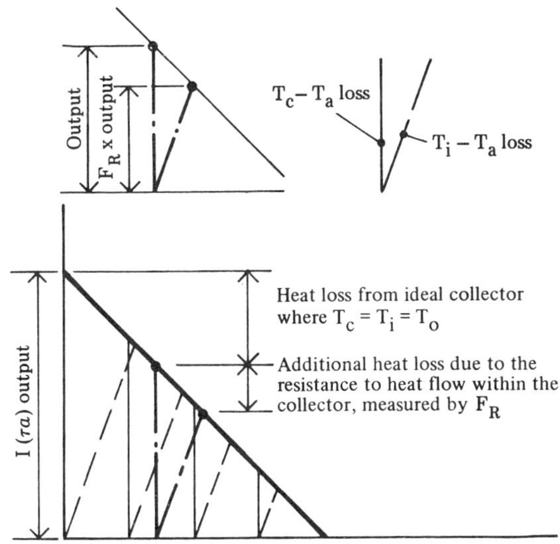

Fig. 4.18 Collector performance graph, superimposing plots based upon average collector temperature and fluid inlet temperature.

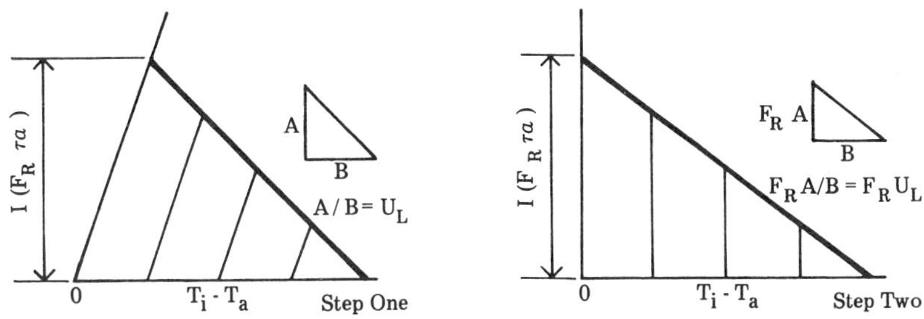

Fig.14.19 The tilted *Y*-axis in Fig. 4.18 can be made vertical, yielding the normal collector performance curve based upon $T_i - T_a$.

Simply eliminating the vertical lines, which are no longer of interest, and "untilting" the graph yields a performance graph based on $T_i - T_a$ (Fig. 4.19). Notice that the heat loss seems to have decreased! This is an artifact of the geometry, and is compensated by a corresponding decrease in e_0. Refer to the overall performance equation:

$$Q = A_C F_R(\tau\alpha I - U_L \Delta T)$$

When it is factored, both $\tau\alpha I$ and $U_L\Delta T$ are multiplied by F_R just as the graph shows. $F_R U_L$ by itself, like $F_R \tau\alpha$, just doesn't have a clear physical

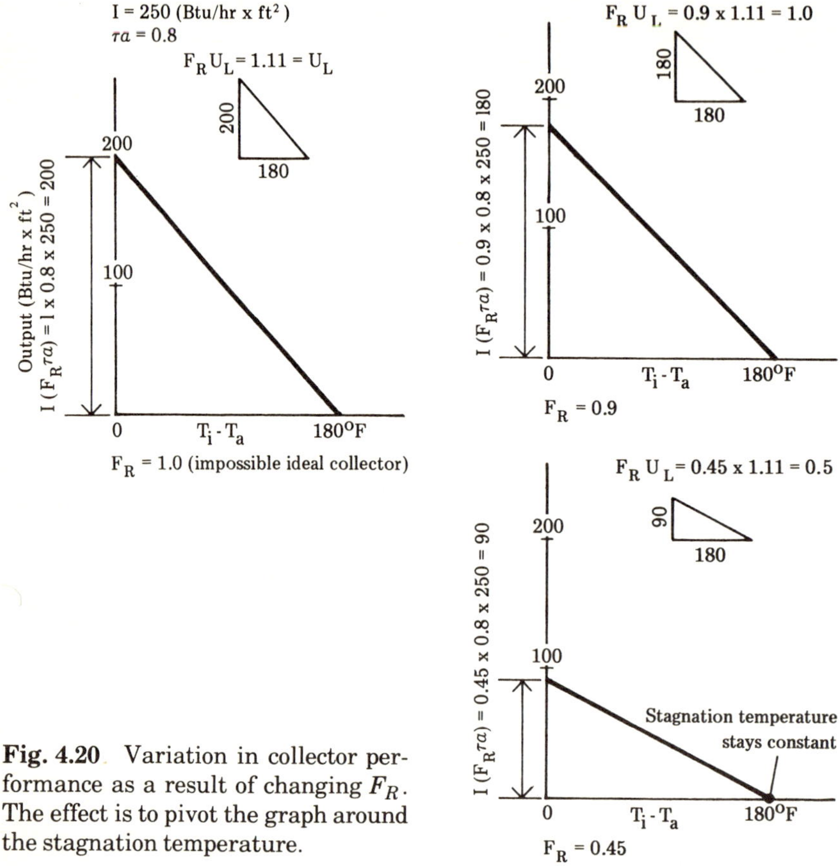

Fig. 4.20 Variation in collector performance as a result of changing F_R. The effect is to pivot the graph around the stagnation temperature.

meaning, whereas $F_R(\tau\alpha I - U_L\Delta T)$ does. Fig. 4.20 shows the overall effect of changing F_R.

Notice that the stagnation temperature is exactly the same in all cases. This stands to reason, since at the stagnation temperature the heat removal factor (F_R) is no longer "working." So we see that raising or lowering the F_R has the effect of rotating the collector output curve around the stagnation temperature point on the X axis.

Now let's fiddle with the U_L by itself. In a single-glazed collector with a painted absorber, the U_L is larger than in the double-glazed collector or one with a selective surface. Let us take the original collector with a U_L of 1.11 and double-glaze it, reducing the U_L to 0.80. Fig. 4.21 shows the resulting curve.

You can see that the overall operation of the collector has improved, especially at high temperatures, and that the theoretical stag-

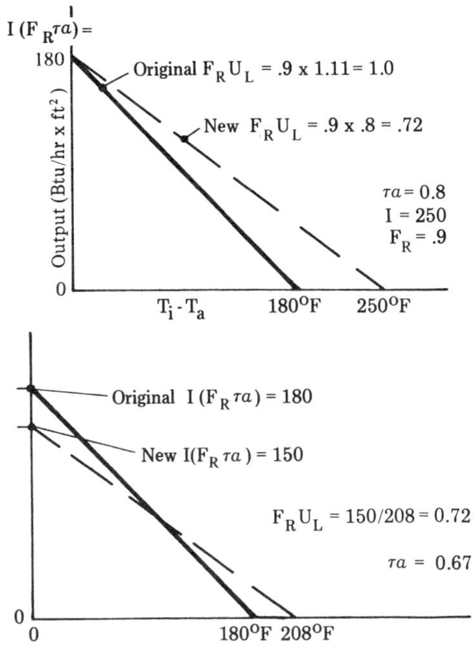

Fig. 4.21 Variation in collector performance as a result of changing U_L. If two glazing sheets are used, the $\tau\alpha$ coefficient usually drops, as shown in the second graph.

nation temperature has gone up from 180°F above ambient temperature to 250°F above ambient. Notice, however, that we have left the tau-alpha component as it was with the single-glazed collector. This means that the two cover layers in the double-glazed collector transmit the same amount of energy as the single cover in the single-glazed collector —a possibility, but one not commonly encountered. More commonly, when the second layer of glazing is added, the $\tau\alpha$ product is reduced along with the U_L, as shown in the second graph in Fig. 4.21. Notice that the stagnation temperature has been reduced.

Another key variable is the rate of insolation, I. To understand how this affects the output, consider two identical collectors, one operating hotter than the other, as the insolation rate increases and decreases during the day (Fig. 4.22). As can be seen, there comes a point with both collectors when the solar energy being supplied just balances the heat being lost to the environment. Typically, collector controls are designed to turn off and turn on this "break-even" point. Notice that the hot collector turns on at a higher rate of insolation than the cooler one. It operates for a shorter period, as a consequence of its higher heat-loss rate. The details of turn-on and turn-off will be examined later in this chapter.

It is worth noting that the balance point is also the stagnation temperature for that particular rate of insolation. Stagnation *problems*

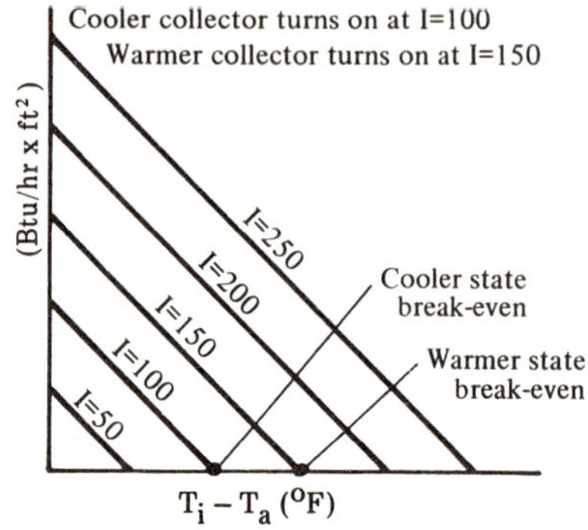

Fig. 4.22 Collector performance graphs for various insolation rates, showing the effect of operating temperature on collector output.

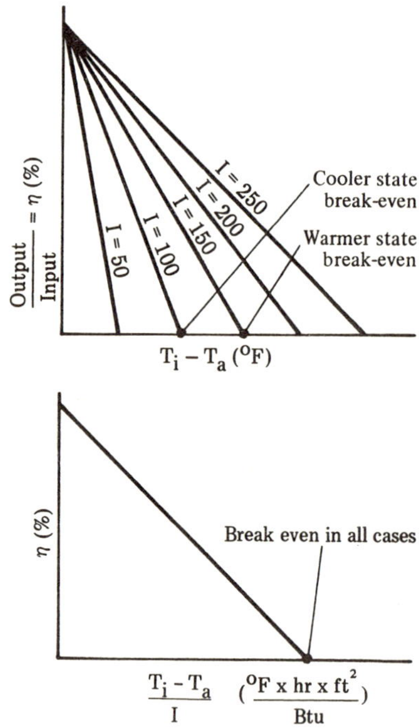

Fig. 4.23 Collector performance graphs transformed by dividing both the ordinate and the abscissa by the insolation rate, normalizing the graph, and making it intuitively obscure.

only occur when the sun is at its peak, the store is so hot that the system must shut down or dump heat, and the outdoor temperature is high.

To establish a bridge between these obvious graphs and the confusing ones used by the industry, first let us convert the output to an efficiency (of sorts) by graphing the output over the input, i.e., Q/I *or* η, for a variety of I's (Fig. 4.23). All the same answers come out of this graph, except that to get the output, you have to multiply η by I. If the ΔT is divided by I, all insolation rates lie on the same line, yielding the convenient and mysterious single line performance curve.

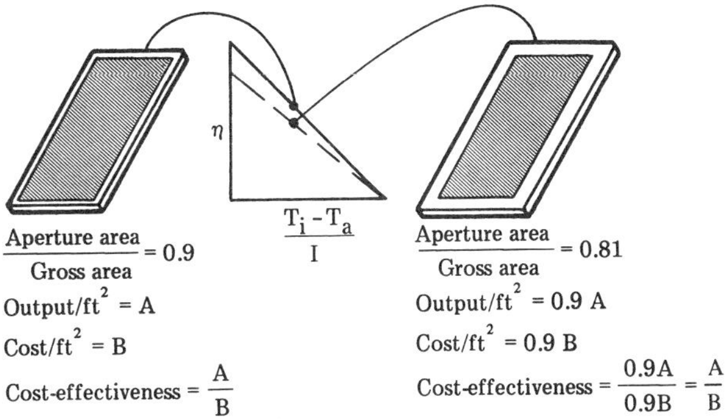

Fig. 4.24 Performance graphs based on collector gross area (ASHRAE test) differ for two otherwise identical collectors with differing ratios of net aperture to gross area. The test effectively encourages a high net-to-gross ratio.

When reading performance graphs, take account of the following:

1. Make sure you know what fluid and what flow rate was used in the test. If low flows with antifreeze were used, you as the designer can improve on the collector output by using higher rates of tapwater, thereby increasing the F_R. The manufacturer can often tell you what percentage F_R changes as the fluid capacitance changes. Some tests are corrected to yield F', requiring the designer to add in the F'' himself.

2. Make sure you know which test was used. The ASHRAE 93-77 uses the *gross* or overall area of the collector to measure the input, although the output depends upon the sun which actually strikes the absorber. The standard makes sense, since it allows one to compare all kinds of collectors with each other—vacuum tubes, trackers, flat plates, etc., and it puts a premium on collectors with a high ratio of net to gross. But it surely confuses the designer.

   Consider Fig. 4.24, graphing two identical collectors, one with a large frame and one with a small one. They cost exactly the same and deliver exactly the same energy to the system, but one looks substantially less cost effective than the other, until you realize that the cost per square foot of the "lower output" collector is commensurately lower, yielding exactly the same cost effectiveness!

3. Although $T_i - T_a$ is now the industry standard by which to measure heat loss, some companies may still use older test methods which involve other collector temperatures, such as $(T_0 + T_i)/2 - T_a$, or even $T_0 - T_a$. Both are attractive, since they, like the net area test, yield "better looking" performance curves. The designer, however, will have great difficulty in using such graphs, since only T_i is easily derivable from system data. Insist that a manufacturer provide data based on T_i.

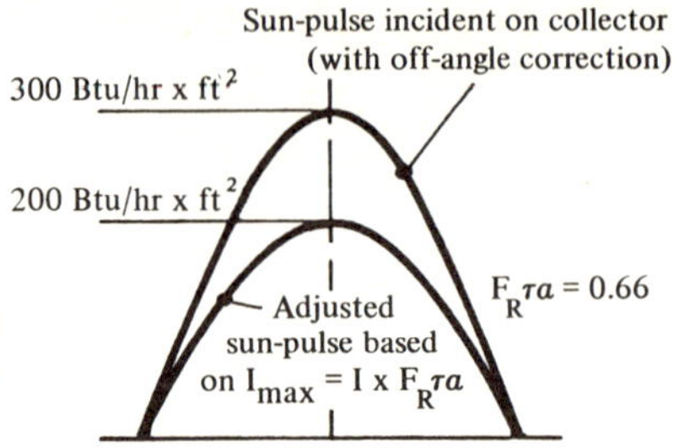

Fig. 4.25 The sun-pulse can be derated to account for the constant $F_R\tau\alpha$ of a given collector.

4. By introducing $\Delta T/I$, all possibility is lost for dealing with the variation in U_L caused by radiation. Consider a $\Delta T/I$ of 0.5. It could represent a hot collector, working at a high ΔT of 150°F but with bright sun at 300 Btu/hr $\times$ ft^2; or it could represent a collector at almost zero ΔT with virtually no sun, just before turn-off. The U_L factor would be different in each case but would have to be plotted as a single line. In effect, one must assume an error band which gets wider as the $\Delta T/I$ gets higher. The alternative is to convert the $\Delta T/I$ graph back to specific graphs for different insolation rates—an exercise heartily recommended for designers.

RELATING COLLECTOR PERFORMANCE TO THE SUN-PULSE

In the preceding sections, in case you missed them, the performance of an active collector was dissected, yielding its two key parameters: $F_R\tau\alpha$ and F_RU_L. It was shown that F_R was to some extent under the control of the designer, while the other factors are published by nearly all manufacturers as the output of the standard ASHRAE performance test. In what follows, it is assumed that the fluid and its flow rate have been set, thereby fixing F_R. The two key parameters can then be assigned fixed values.

We saw that, despite the intrusion of F_R, the first parameter dealt with the efficiency of the collector as an optical device, transforming incoming high-frequency radiation into low-frequency heat within the absorber. The other parameter dealt solely with the loss rate and could be considered along with the other losses from the system. Because $F_R\tau\alpha$ is a constant which derates the incoming radiation, it can be applied directly to the sun-pulse, reducing its amplitude (but not changing its shape or duration). For instance, if the sun-pulse is defined as a sine curve with a peak amplitude of 300 Btu/hr $\times$ ft^2, and if the collector's $F_R\tau\alpha$ is 0.70, a new sun-pulse can be used with a peak amplitude of 210 Btu/hr $\times$ ft^2. We then have to deal only with heat losses (see Fig. 4.25). As long as we deal with a unit area of collector, it is possible to express

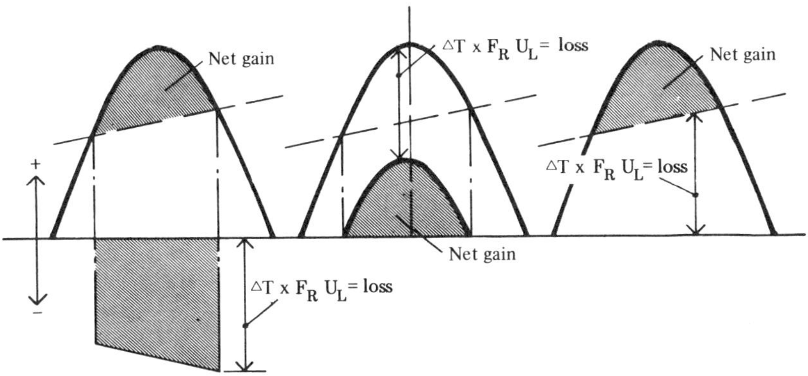

Fig. 4.26 Three ways to plot collector losses on the sun-pulse curve in order to determine the net gain. The method on the right will be used in this book.

the instantaneous heat loss at the same scale as the gain, represented by the sun-pulse amplitude. Three convenient ways to diagram this are shown in Fig. 4.26.

In the first example, the losses are plotted below the zero line, with the sun-pulse gain plotted above it. At the two vertical lines where the gain equals the loss, the collector will turn on (neglecting the controller deadband). The net output will be the total gain between turn-on and turn-off, minus the loss. In the second example, the loss is subtracted vertically from the gain, yielding a misshapen sine curve lower than the sun-pulse. The part of this curve rising above the zero line outlines the net gain. The third superimposes the loss and gain curves. All three diagrams yield exactly the same result. The first is easier to use in calculating, the last in visualizing what is going on. We will use the last form in the following discussion.

Let us examine various changes in the collector one by one. First, compare a collector with a low $F_R\tau\alpha$ to one with a high $F_R\tau\alpha$, with the heat loss constant (Fig. 4.27). On a cloudy day, the difference is crucial: the collector does not turn on. Now reduce the heat loss for the three examples (Fig. 4.28).

The heat-loss reduction can be achieved either by reducing the *collector-loss rate* or by reducing the *operating temperature*. If simple, homemade or plastic collectors are to be used, it is essential to reduce the operating temperature, since there is no way to decrease the heat loss. (Remember—decreasing the heat loss increases stagnation temperature, requiring more sophisticated materials, and perhaps making simple construction impossible.)

If the collector is very massive, some of the energy from the sun-pulse will be used to raise the temperature of the collector and its

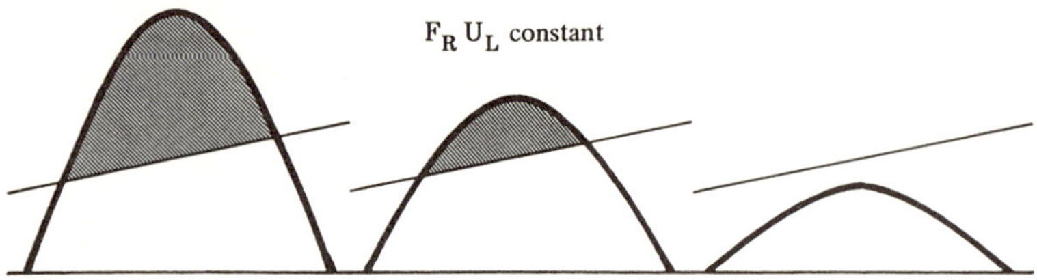

Fig. 4.27 A reduction in $F_R\tau\alpha$ causes a disproportionate reduction in output when the loss rate nearly equals the available insolation.

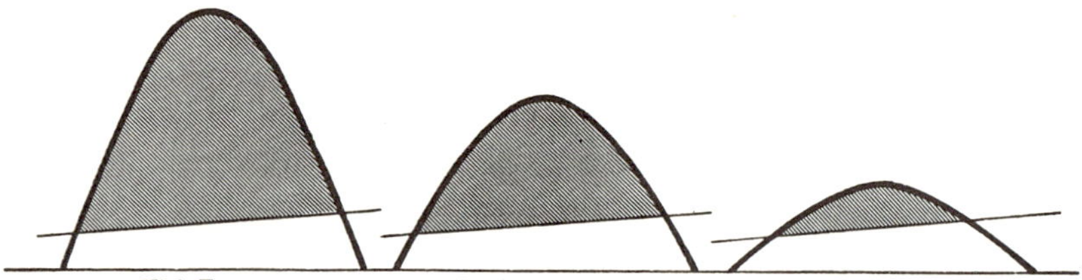

Fig. 4.28 At lower loss rates relative to available insolation, changes in $F_R\tau\alpha$ are commensurately less serious.

contents. Normally, this energy is given back when the collector cools. In abruptly changing sunlight, excessive mass may prevent turn-on (see discussion in this chapter concerning controls).

A CAUTION ABOUT AVERAGING

In the use of sun-pulses, it is tempting to keep simplifying, perhaps to the point of gross inaccuracy. In the diagrams below, a type of inaccuracy is shown which results from "overaveraging." Suppose you decided to model two sun-pulses, one 50 percent of the other, by using two sun-pulses each with 75 percent the value of the first. In Fig. 4.29, you would use two "B" sun-pulses to replace the "A" and "C" sun-pulse. Then, suppose the heat-loss rate was such that it exactly equaled the amplitude of the "B" sun-pulse (this could happen with a hot collector).

It can be seen that the net gain using the "B" sun-pulse is zero, whereas "A" + "C" yields the "A" gain. As with other simplifications in the sun-pulse method, the error is conservative and occurs at low insolation levels. On normal sunny days, the loss line is so far below the peak of the sun-pulse curves that the error is negligible, as in the second graph in Fig. 4.29.

A similar error can occur when using averaged ambient tempera-

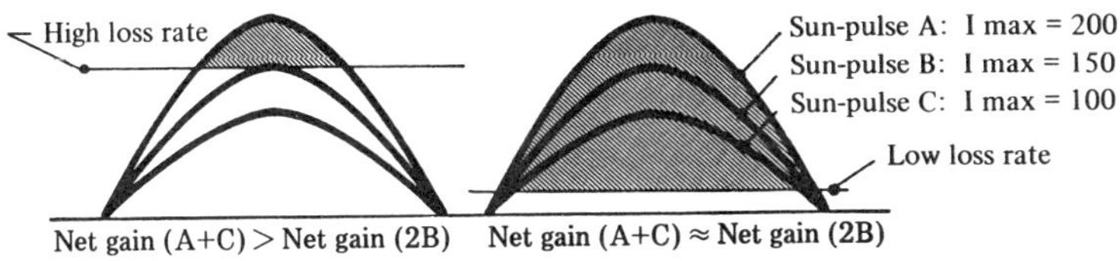

Fig. 4.29 Caution is called for in averaging insolation rates when running simulations. If the loss rate is close to the average maximum available, insolation, as in the left-hand example, serious errors can result if averages are used.

tures, again an effect noticeable only at low insolation levels. In both cases, the antidote is to "exercise" the system over a range of levels, rather than sticking strictly to averages.

CONTROLLING ACTIVE COLLECTORS

It should now be evident to the reader that the performance of active collectors is improved by increasing the rate at which the working fluid can withdraw heat from the collector. The cooler the collector, the more heat passes into the fluid. When we ask how to turn such a system on and off, many questions arise, and careful analysis is necessary in order to understand the complexities of the problem.

Put succinctly, the collector operates in two modes: Before the fluid begins to move, the collector has a relatively low loss rate, and so rises rapidly in temperature as the insolation rate increases. With the fluid flowing, especially in a liquid-cooled collector with a high rate of flow, the "loss rate"—that is the rate of heat flow from the collector—rises dramatically, along with the amount of insolation needed to raise its temperature. At turn-on and turn-off this abrupt discontinuity in the temperature response to the insolation rate puts severe constraints upon the design of a control system.

Almost universally, a device called a "differential controller" is used to control solar collectors. This device measures the difference in temperature between two sensors by comparing the resistance to current flowing through the sensors, which are designed to exhibit a linear relationship between their temperature and their resistance. One sensor measures the temperature at the bottom of the storage tank, at the point where the fluid returns to the collectors, while the other measures the temperature at the collectors themselves. It is not uncommon for the collector sensor to be placed on the outlet pipe from the collector, as close as possible to the collector plate, and insulated from the environment. In this location, the sensor alternately measures the condition of the collector in its two operating modes—first stagnant, then with the fluid flowing. Such a control system used in a high flow rate liquid-cooled collector will always "cycle," or repeatedly turn on and off, during start-up and shut-down. In the worst cases, such systems can cycle all day long.

Contrast this with a collector sensor mounted in a little collector all by itself. This little collector is designed to have a thermal response to insolation identical to that of the operating collectors, except that it never has fluid flowing through it. Now the sensor will operate the system correctly, since it responds only to one operating mode of the collectors, and the one (important to note) which exhibits the largest temperature swings. By measuring large swings, the inherent errors in the sensors and controller become minor. If the fluid temperature changes were to be measured, fractions of a degree would have to be sensed, requiring very sensitive and expensive equipment. The recommendation, therefore, is to isolate the collector sensor as much as possible from the influence of the fluid passing through the collector. This is a good strategy on any collector, but becomes essential for collectors with high flow rates.

Since this area of solar design has received more than its share of obfuscation, it is worthwhile to examine the problem in some mathematical detail, and derive expressions for the difference in thermal response for the two modes as a function of parameters over which we have control in designing systems.

In its first mode, with no fluid flowing, the collector is in stagnation. All the energy that passes into the collector must be dissipated into the surrounding environment, so that:

$$I(\tau\alpha) = (T_{\text{collector}} - T_{\text{ambient}})U_L$$

If we assign T_{ambient} the value of zero, and measure all temperatures from that baseline, we can write:

$$T_{\text{collector}} = I\,\frac{\tau\alpha}{U_L}$$

This makes sense. Collector temperature is linearly proportional to insolation rate (as seen in a linear collector performance plot); and the higher the loss rate and the lower the $\tau\alpha$ product, the more insolation is required to raise the collector temperature.

For example, if $\tau\alpha = 0.75$ and $U_L = 1.25$, each Btu/hr $\times$ ft^2 entering the cover will raise the collector 0.6°F ($0.75/1.25 = 0.6$).

Because stagnation temperature is linear with insolation (as long as we stay at reasonable temperatures and do not wander out onto that portion of the performance curve where radiation losses take over), we can convert an ordinary sun-pulse curve into a stagnation temperature curve (Fig. 4.30).

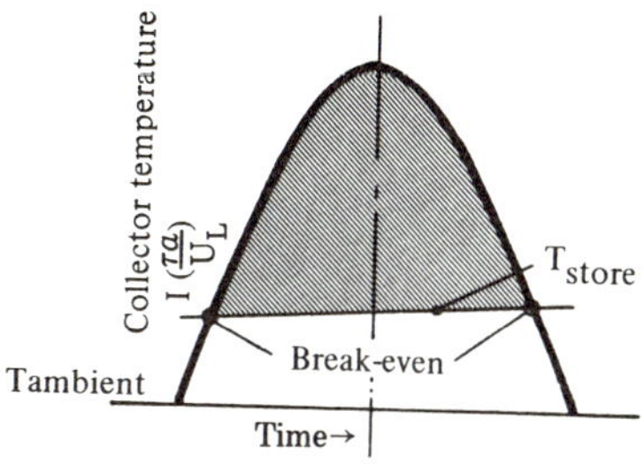

Fig. 4.30 A sun-pulse curve can be converted into a collector stagnation temperature curve, if a linear performance curve is assumed. The break-even condition occurs when the collector temperature equals the store temperature. The diagram is exactly like that in Fig. 4.26, except that temperature and not power is the vertical axis.

On the diagram, a "break-even" line is shown, describing the condition at which the collector temperature is exactly equal to the temperature at the bottom of the storage tank, where the second control sensor is located in a good design. (Be sure to keep sensors outside the tank, under the insulation. Wells tend to fill with condensation and cause incorrect readings, such as −50°F!) It is the goal of control design to turn the collectors on and off as close as practical to this break-even condition. Note that at the break-even condition, if we turn on the collector pump the fluid will enter and leave the collector at the temperature of the plate. The temperature of the absorber is the same whether or not the fluid is flowing. In general, running a fluid through a collector at the current stagnation temperature has no effect on either the temperature of the fluid or that of the collector.

In its second mode, when fluid flows, the collector equation yields a more complex relationship between insolation rate and collector temperature. The basic equation states that all the energy coming in must go either to the fluid, or to the environment, in proportion to their heat-removal rates. However, the fluid temperatures in turn influence the loss rate to the environment. The equation can be written like this:

$$T_0 - T_i = \frac{IF_R\tau\alpha - F_RU_L\left(\frac{(T_0 + T_i)}{2} - T_a\right)}{Gc_p}$$

where T_0 = Temp. Fluid out

T_i = Temp. Fluid in

G = Mass flow rate of fluid per ft^2 of collector

c_p = Specific heat of fluid

$\frac{T_0 + T_i}{2}$ = Approximate average temperature of collector

In this form, it states that the net heat after losses to the environment raises the fluid temperature proportionally to the "capacitance rate" (mass flow rate per unit area × specific heat) of the fluid. The proportionality relationship is more conveniently written relating three

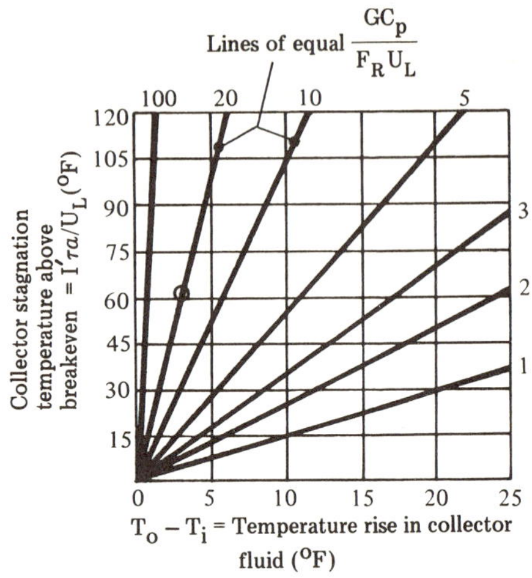

Fig. 4.31 Together, the collector loss rate ($F_R U_L$) and the fluid capacitance rate (Gc_p) determine the rise in the fluid temperature through a collector, for a given amount of insolation. In this graph, the quotient $Gc_p/F_R U_L$ is plotted against the collector stagnation temperature, measured above the break-even condition, to yield the expected rise in fluid temperature if the collector were running at that point. The chart predicts how hot the collector must be above T_i to prevent cycling, given a collector sensor which measures fluid outlet temperature. The example condition described in the text is shown by the small circle.

terms:

$$I' \times \frac{\tau\alpha}{U_L} \qquad \frac{Gc_p}{F_R U_L}; \qquad \text{and} \qquad T_0 - T_i$$

where I' is the insolation in addition to that required to bring the collector up to temperature T_i (break-even). The equation is:

$$T_0 - T_i = \frac{I'(\tau\alpha/U_L)}{(Gc_p/F_R U_L) + 0.5}$$

This form of the equation exposes the roles of the capacitance rate and the collector loss rate, and uses the same multiplier for I' that the equations for the first operating mode used for I. It is possible to graph the relation among the three terms (Fig. 4.31).

The graph can be used to predict directly the influence of the fluid capacitance rate on collector response. For instance, if the capacitance rate has the ratio of 20 to 1 relative to the collector heat loss rate, the collector in the previous example would require an insolation rate of nearly 103 Btu/hr × ft^2 (corresponding to a rise in plate temperature of 62°F) above the break-even condition to raise the fluid leaving the collector 3°F above that entering it.

The practical consequences of all this relate to the control process. Since some sensors may exhibit errors of more than 1°F, it is necessary to set the turn-off differential *above* the break-even point, in order to guarantee that the collector turns off before the fluid starts to lose heat. If the turn-off differential is 3°F (a good minimum), and if the collector sensor is mounted onto the water pipe at the collector, a turn-on differential of nearly 70°F would be required to prevent cycling (in the example above). Such a large differential would keep the collector off most of the year; typically, controllers are equipped with 15°F to 20°F turn-on differentials.

Looking at the chart, it is easy to see that as the flow rate is reduced, a smaller and smaller insolation rate, and therefore turn-on differential, is needed to avoid cycling (given the collector sensor under the direct influence of the fluid stream). Such a strategy runs directly counter to sound thermodynamics, and must be rejected in favor of the more sensible approach of disassociating the sensor from the fluid.

In this preferred arrangement, the collector sensor registers the *potential* for the collector to add heat to the fluid stream, but does not try to measure the actual rise in temperature, which is really too small to measure. A similar problem arises in swimming pool heaters, where a large flow of water is often used. Since the collectors are unglazed, and since they pick up a large part of their heat from the air, the sensor is attached to a small, separate, flat-plate collector with absorbing characteristics similar to those of the actual collector. In a glazed collector, the same thing can be accomplished by cutting a piece from the absorber, attaching the sensor firmly to it, and resetting the detached piece in place, having cut it a little smaller than the hole it left.

Another problem with sensors attached to the piping can be solved by a separated sensor. If for some reason the tank is very cool, and the temperature of the fluid leaving the collector drops below that of the entering fluid due to an inadequate turn-off differential, it is theoretically possible for a system to continue running in the absence of the sun, since the fluid moving through the collector supports the temperature at the attached collector sensor and prevents it from shutting the system down. By having the sensor register only the insolation and collector loss, a positive shut-off is guaranteed.

One reason for a turn-off differential of 3°F or even 5°F is to compensate for lightweight absorber plates, which with their short time-constants, can cause systems to cycle rapidly in "pop-corn" cloud weather. If this is considered a disadvantage (as it should be in drain-back systems), the sensor can be mounted to a heavier plate than that of the remainder of the absorber, providing it has the same coating. The heavier plate will make the "sensor collector" have a longer time constant.

As long as the small absorber for the sensor is in the same box as the collector itself, it will sense the temperature of the air between the absorber and the glazing. This air is influenced by the overall absorber-plate temperature, which is in turn affected by the fluid temperature. This short-circuit will to some extent counteract the advantages of a separate sensor, but the effect is usually minimal (except, of course, when the sensor is in the air stream of an air collector). In addition, the stagnant air at the top of the collector, where the sensor is normally placed, is usually hotter than that at the bottom, resulting in premature turn-on and late turn-off. (This is true whether the sensor is separated or not.) But remember that except for rare occasions, the collector does not stay at the break-even point very long. Usually it passes through the break-even condition rapidly, resulting in a decisive turn-on or turn-off instruction (with a separated sensor) even if a short period of negative gain is experienced.

It is important to note that between the two extremes of sensor location discussed above, intermediate solutions are possible. Perhaps the most common is to locate the sensor on the front or back of the absorber, as far from the fluid passage as possible. This introduces the temperature lag between the fluid stream and the hottest part of the absorber into the equation, so that the sensor gets a mixed signal. One part is linearly proportional to the insolation rate; the other is proportional to the fluid temperature. The trouble with this strategy is that it is often inadequate. To make it work, that is, to prevent cycling, one must either reduce the fluid flow rate or reduce the fin efficiency of the absorber, or both. But both of these reduce collection efficiency. Relying upon sub-optimum thermal design to solve the control problem seems to the author counterproductive.

A careful mathematical analysis of the intermediate case is beyond the fragile resources of the author, and probably would try what is left of the reader's patience. Some of the newer technical texts will provide the necessary formulas.

Another point which needs making is that air systems are far less sensitive to the control problems described, because the capacitance rate is close to the collector loss rate (typically 2:1 or less, as opposed to 20:1 in a water cooled collector). The effect is evident in Fig. 4.31.

GUIDELINES FOR DESIGNERS

1. Because of the great difference in thermal response between a collector in stagnation and one being cooled by a liquid running at a high flow rate—for instance, 1/2 GPM of water per 20 ft^2 collector module or higher—it is recommended that the collector sensor be mounted on a piece of absorber plate separated from the rest of the absorber. Otherwise, such a system will cycle unless an unreasonably high turn-on differential is used.

2. The same conclusion can be reached in trying positively to prevent any tendency for the system to run with a negative turn-off differential, and thereby risk never turning off. With a sensor on a separate plate, positive turn-off is guaranteed.

3. A 3°F to 5°F turn-off differential is essential to compensate for sensor and controller error. This also compensates somewhat for very long and/or poorly-insulated piping runs, which cause a slight reduction in fluid temperature, and may cause losses at shut-down.

4. While it is theoretically advantageous to minimize the turn-on differential, it should be enough higher than the turn-off setting to prevent rapid cycling during intermittent clouds. A good combination might be 5°F and 15°F. The small loss at turn-on is not nearly as important as smooth control and positive shut-down.

5. Although not discussed above, proportional control is sometimes used, despite the cost of the pumps and controls. It normally cannot be used with a drain-back system, as will be explained elsewhere, because of the necessity to form a syphon. In other systems, it artificially holds the operating temperature of the system higher than necessary at start-up and shut-down. The strategy recommended above should solve most control problems, and obviate the need for proportional control. Notice that proportional control adjusts the ratio of flow rate to collector loss, rather than dealing directly with the sensor problem. Again, good thermal design should not be sacrificed to solve a control problem.

6. If a substantial amount of shading occurs on the collector array, a special analysis should be made of the control system, to prevent extensive operation below the break-even point. Additional collector sensors may be necessary. (Consult a reputable control manufacturer.)

COLLECTOR CHECKLIST FOR DESIGNERS:

Test

(Modular) ASHRAE 93? uses $T_i - T_a$

Others? may use $(T_i + T_0)/2 - T_a$ and/or net area

Certified lab? (which one)

What ambient and wind conditions?

What fluid capacitance, what fluid?

Is collector when installed subject to same conditions as test (i.e., is back protected or exposed; are collectors mounted together)?

Same collector tests as specified?

Passed stagnation?

Severe off-angle drop-off?

Manufacturer

Reputable?

How long in business?

Offer HUD 5-year warranty?

Markets in your area?

Reliable reps?

History of good workmanship?

Handle other products in construction industry (likely to continue service for good will)?

Shell Construction

Frame Material—aluminum (problem attaching to steel), steel (may rust—needs galvanizing), wood (must be flashed, and pressure treated or durable species)

Self-supporting? Extra strength to help out in supporting array?

How fastened? Compatible with normal details? End flanges, Edge detail, Back connected, No provision (can be costly)

Flashed in? If so, how do intersections work? If not, how does water get out, air flow behind in standard detail?

Glazing well secured?

Will frame distort when fastened down?

Sealant durable? Joints caulked?

How do collectors join each other?

Weight: difficult to lift?

Size—compatible with many installations?

How much room needed to manifold?

Absorber—Liquid

Internal or external manifold? (internal may save cost)

Drainable without tilting collector?

1″ or 3/4″ manifold (if internal)? 1″ allows good water flow

Do drain-downs have corner-to-corner outlets? (good for fluid flow, bad for stacking vertically, since outlet not above inlet)

Separate fluid passages? If so, how are tubes attached to plate or fins—good joint?

How far apart? Is fin "*kt*" sufficient for spacing?

Tube material: If aluminum, how is fluid treated to avoid corrosion? If steel, no oxygen can be allowed in system. Copper is standard. If nonmetallic, is performance satisfactory?

How much can the absorber move? (affects number of collectors which can be soldered together)

How are collectors joined to piping? Is there room? Will soldering or clamping hurt the casing?

Absorber—Air

Is absorber excessively heavy?

Is manifolding simple? Are there minimal number of joints?

Is air in a sealed duct? (leakage otherwise)

Does absorber have fins? (improves performance)

Coatings

Paint? Durable?

Selective surface? How good ($\epsilon = 0.25$ to 0.10)? Professionally applied? Properly primed?

Glazing

Glass or plastic?

If glass, is it tempered? (should be) Nonreflective? (helps off-angle loss) Low iron? (better transmission)

If plastic, what kind?

Acrylic (Plexiglas, Acrylite)—expensive, moves a lot, durable, very transparent.

Polycarbonate (Lexan)—very durable, very expensive, moves a lot, transparent.

Polyethylene—unsatisfactory (U.V., strength).

Tedlar film—very tough; by itself degrades due to heat. With Teflon inner film, may be okay. Heat shrinks, then expands and contracts a lot.

Teflon film—heat resistant; good inner layer, not available thick or strong enough for outer layer.

Fiberglass reinforced polyester resin—cheap, withstands moderate temperatures, strong and stiff, moderately transparent.

If plastic, how sealed? (most common problem with stiff plastics —not enough sealant)

If film, how secured?

Double-glazed with selective surface? (waste of money)

Properly vented? (at bottom, to prevent condensation)

Insulation

Adequate for installation intended? (Remember, if roof mounted, may be on stand-off; if integral, may have vent space below but no wind.)

Protected from moisture and abrasion?

If polystyrene, is it up against absorber? (won't take the heat)

If fiberglass, is it either not subject to temperatures over 350°F, or oven dried? (to prevent outgassing)

Site-Built and Trickle-Down

Is water exposed to air below glazing? (condensation may be a problem)

Is absorber able to stand up with water running down coated surface?

Is water in contact with raw aluminum?

Is wood or other fibrous material in the air stream? (Some woods are reported to ignite if repeatedly heated over 250°F; other materials can be abraded by moving air.)

Can parts move adequately relative to each other under thermal stress?

Is glazing sealed properly?

Is the space below a built-in collector safe in case of leakage or blow-out? (best to mount collector above roof sheathing)

5

HEAT EXCHANGE HEATING SYSTEM

BASIC CHARACTERISTICS OF HEAT EXCHANGERS

A building, considered as a thermal artifact, can be thought of as a membrane separating two volumes of fluid which are commonly at different temperatures. As such, it falls into the large category of things known as heat exchangers. In addition to the building itself, there are likely to be a number of heat exchangers within the building—people, ovens, refrigerators—along with specialized devices *called* heat exchangers. There is a heat exchanger in every furnace and boiler. Duct coils and baseboard convectors, radiators, fan-coil units and instantaneous heaters are all forms of heat exchangers.

In order to simplify our understanding and allow the use of simple computer models in solar heating design, it is essential that we express the performance of all these heat exchangers in the same dimensions. These dimensions are those we use to define the heat loss of a building. Although not precisely true, due to the influence of radiation, ground conditions, stratification and infiltration, it is approximately true that the *rate* at which heat passes from the inside of a building to the outdoors is proportional to the temperature difference between indoors and out. We can, and do, define a proportionality constant relating the rate to the temperature difference, called the "U-factor." The U-factor, multiplied by the area of the membrane, results in a "UA" value with dimensions Btu/hr × °F. The UA values for all the building surfaces can be summed into a term "ΣUA." A 1400 ft^2 house just meeting minimum standards in 1979 would have a ΣUA loss rate of about 500 Btu/hr × °F. One can multiply this value by 24, yielding a heat loss rate per day per °F, a useful number when dealing with the building's annual heat loss.

Contrary to our usual practice of trying to make a building a very poor heat exchanger, suppose we instead try to enhance heat transfer through its walls. First, we would redesign the building to maximize its perimeter. We might, for instance, create a long thin floor plan, and add lots of fin-like projections. Next we would replace the insulated walls with thin sheets of thermally-conductive material—copper, perhaps. Most of the remaining insulation would now be in the form of surface films of still air on each side of the membrane. If we could scrub off these films by moving the air more vigorously across the surface of the membrane, we could greatly increase the *UA* value. For instance, our copper wall might have a *UA* value as low as 0.7 with still air on both sides. By whistling up a brisk wind outside, and directing a fan on the inside surface, we might get the *UA* loss rate up to 3.0 or more.

The methods used can be generalized:

- Increase the heat transfer surface area of the membrane.
- Make the membrane wall as thin as practical.
- Make the membrane from a material with good thermal conductivity.
- Scrub away the surface layer of fluid on both sides of the membrane.

Enhancing leakage across the membrane, a method used extensively by builders to increase building heat loss (infiltration), is usually inadvisable, dangerous, and/or unlawful when applied to heat exchangers. The heat exchanger is there in the first place because the two fluids could not be mixed together. If you *can* mix them, as the collector fluid, store fluid, and heating fluid are mixed in our prototypical drainback system, don't add a heat exchanger if you can help it.

Although the design of a heat exchanger can be completely described by listing its *UA* values for various fluids and flow rates, one must know what these latter are in order to determine the actual output. One knows that air is present both inside and outside a building, and that there is for all practical purposes no limit to the quantity of air. In a packaged heat exchanger, however, only a limited amount of fluid can be run through each side in a unit of time, and this fluid may vary in its ability to carry heat. One parameter defines both these characteristics: the "heat capacity rate," which is the product of the mass which flows per unit of time, and the specific heat of that mass. In symbols, one can use $\dot{m}c_p$, or more conveniently, $\dot{C}$. (The term *Gep*, used in Chapter 4, is *C* per unit of collector area.) $\dot{C}$ has the dimensions Btu/hr × °F, the same as *UA* (or *AU* as it is called when dealing with heat exchangers). In a sense, *AU* represents the membrane's capacity for transmitting heat, while $\dot{C}$ represents the fluid's capacity for supplying heat to the membrane. The two are to some extent interactive. A higher $\dot{C}$ can be the result of an increased flow rate, which in turn will scrub more of the surface film layer from the membrane and raise the U factor. Most of the mathematical complexities surrounding heat exchangers involve methods of calculating the film coefficient "h" as the viscosity, density, velocity, thermal conductivity, and specific heat of the fluid(s) change, for given geometries and given relationships with gravity or other external forces. While one might like to understand Reynolds, Prandtl, Nusselt, and Grashof numbers, most practicing designers will have to forego this pleasure and rely instead upon approximations, data supplied by manufacturers, and, occasionally, experimentation through his designs.

One can generalize to some extent:

- In forced-flow heat exchangers, the AU value doesn't change much over the range of flow rates and temperatures involved in normal operation.
- In gravity flow heat exchangers, important changes take place in the AU value as the fluids change temperature and/or velocity. Sometimes these changes are confined to low operating temperatures, as in the baseboard finned-tube convector, which stabilizes into a more or less linear performance curve after the water rises about 20°F higher than the air in the space.
- It is pretty silly to calculate the mathematical characteristics of a heat exchanger precisely, while basing the design upon the wrong premises. It is almost never useful to know exactly what a heat exchanger is doing in a particular state of a solar heating system, as long as you have made it reasonably cost-effective. And the amount of information you need to assess cost-effectiveness is quite small.

DESIGNING WITH HEAT EXCHANGERS

Heat exchangers can be considered to be 100% efficient, since all the energy which goes in one end comes out the other. This is obviously not precisely true, but losses from the device can be considered along with the other losses in the system; its performance and its efficiency can be considered separately without much error. A new term must be introduced to describe how well or poorly a heat exchanger works. While "well-insulated" or "poorly-insulated" would do, a more precise term, "effectiveness," is used. Effectiveness is defined as the ratio of the *actual* rate of heat transfer through the exchanger to the *theoretical maximum* rate.

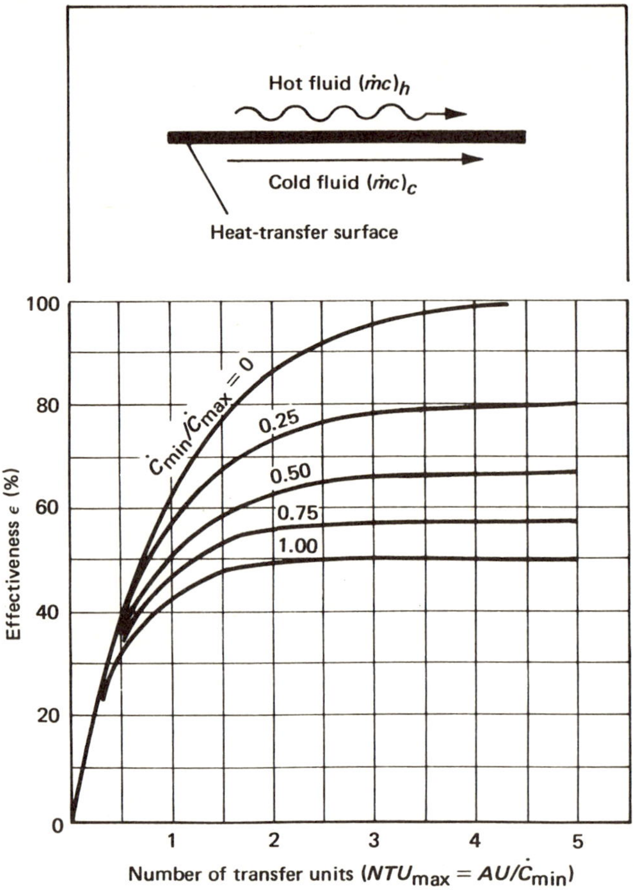

Fig. 5.1 Effectiveness for parallel-flow heat exchangers. This and the following five figures plot heat exchanger effectiveness as a function of the quotient $AU/\dot{C}_{min}$, the maximum Number of Transfer Units (NTU_{max}). The drawings are reproduced from Kreith and Kreider, "Principles of Solar Engineering," 1978, Hemisphere/McGraw-Hill. The original charts are found in Kays and London, "Compact Heat Exchangers," 2nd ed., McGraw-Hill Book Co., NY, 1964.

In the charts reproduced in Figs. 5.1 through 5.6, the effectiveness of several kinds of heat exchangers is given as a function of the dimensionless parameter formed by dividing UA by the lesser of the two $\dot{C}$ values ($\dot{C}_{min}$), called NTU (number of transfer units). Before considering the theory which leads to these charts, the various types of heat exchangers need a brief explanation.

Many packaged heat exchangers come in the "shell and tube" configuration, which is essentially a bundle of pipes (the tubes) con-

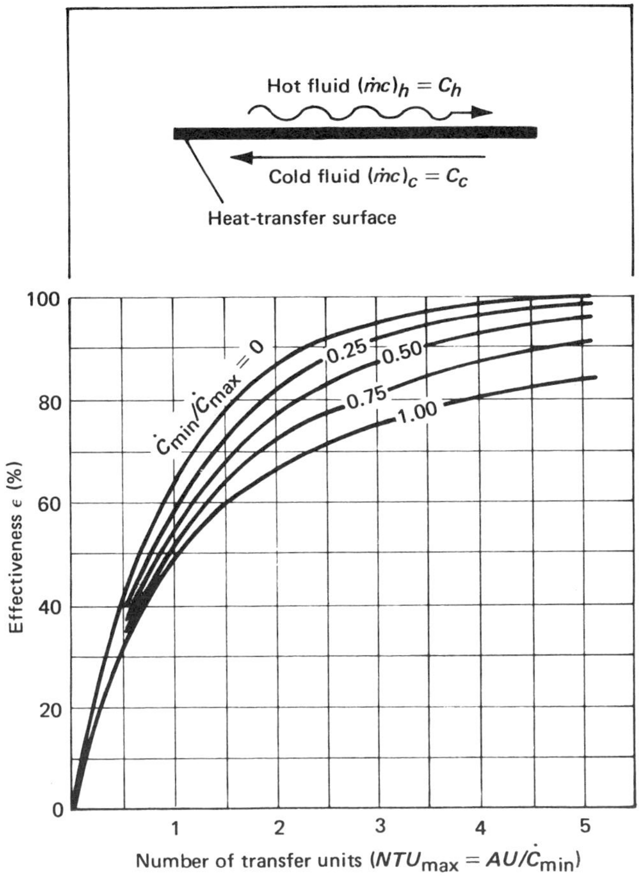

Fig. 5.2 Effectiveness for counterflow heat exchangers.

tained in a larger pipe (the shell). Typically, the tubes pass through an impervious "tube sheet" at each end, outside of which is a chamber called a "header." Fluid can enter one header, pass through the tubes, and exit from the other; or the headers can be partitioned, forcing the tube fluid to pass back and forth from two to eight times. Similarly, the shell fluid can enter one end of the shell and exit from the other; or it can be forced back and forth a number of times. In most designs the shell fluid is forced around a cleverly-designed series of baffles to pass it evenly across the tubes and to increase its velocity. While the tubes are generally copper and the tube sheets brass, the shell and header can be a variety of materials. When handling fluids with oxygen in them, such as tapwater, it is essential that the oxygenated fluid contact only cuprous metals or stainless steel. Bronze, brass, and copper heat exchangers are

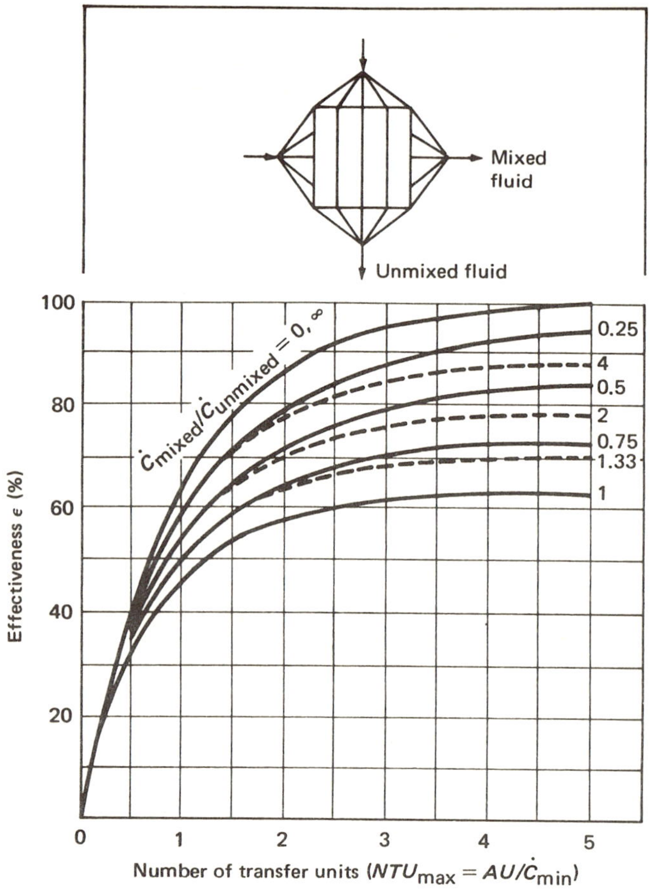

Fig. 5.3 Effectiveness for cross-flow heat exchangers with one fluid mixed.

much more expensive than those with steel shells and cast-iron headers, a factor which helps enhance the attractiveness of immersion coils in drain-back systems.

The various types listed in Figs. 5.1 through 5.6 are:

Counterflow–Two liquids are forced to pass through a small diameter heat exchanger in opposite directions. This design tends to be more costly, but also more effective. It should be considered when the flow rates on both sides are roughly equal.

Parallel Flow–Similar, but with the flows in the same direction.

2-, 4-, 6-, and 8-Pass, and Multi-Pass–Most heat exchangers, for reasons of cost and compactness, are combinations of the above two

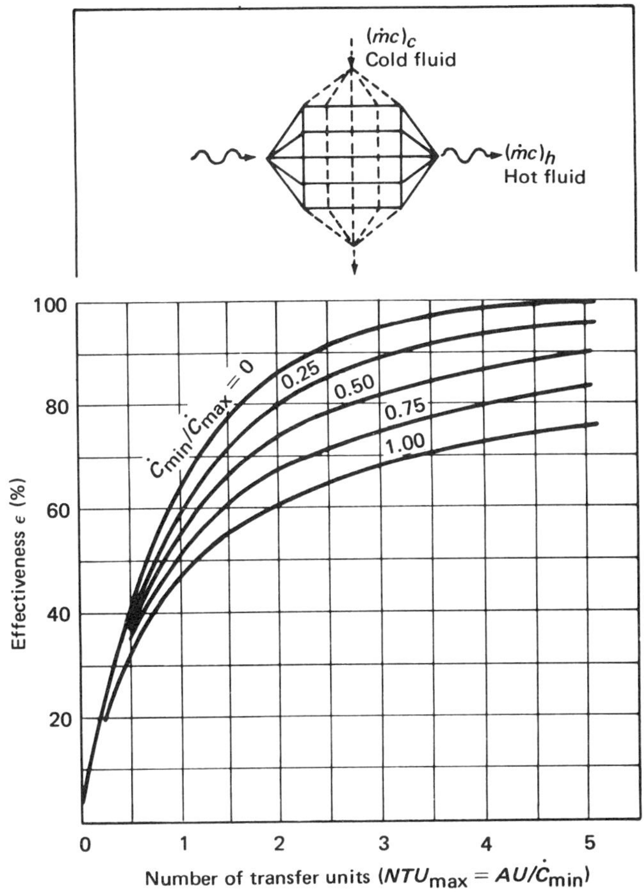

Fig. 5.4 Effectiveness for cross-flow heat exchangers with both fluids mixed.

types. In a 1-shell pass, 2-tube pass design, for instance, the header at end A is split, with the tube fluid entering one side, passing through half the tubes, and exiting from the other side at end A. The shell fluid enters at end B and passes through the shell to end A. In a 2-shell pass, 4-tube pass design (multi-pass), the shell fluid enters and exits from the same end, being forced to pass down one baffled chamber and back another; while the tube fluid enters one quadrant of header A, turns around through two quadrants of header B, turns around again through two quadrants at end A, back through the remaining quadrants at end B, and finally exits from the remaining quadrant at end A, having at each pass traveled through a different one-fourth of the available tubes.

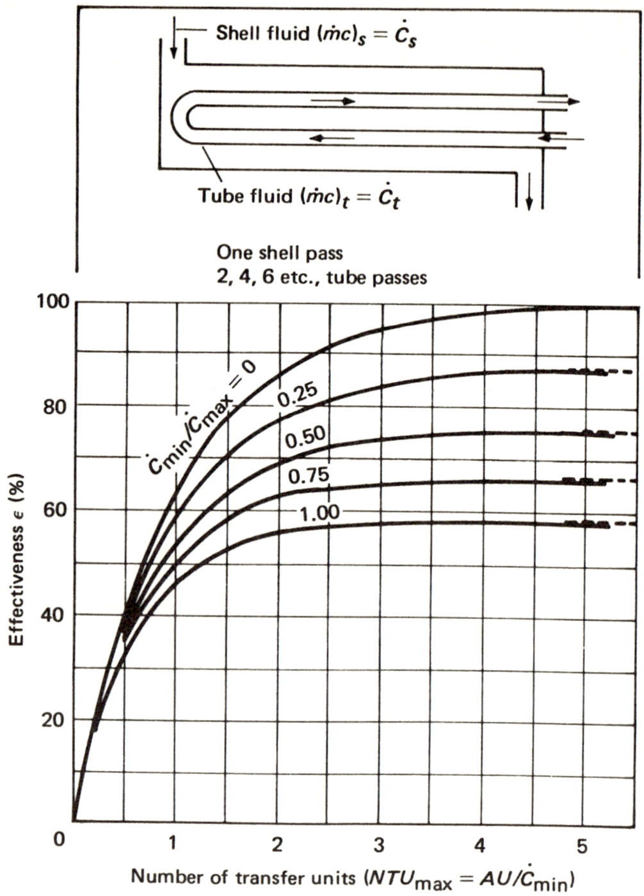

Fig. 5.5 Effectiveness for heat exchangers with one shell pass and two tube passes.

Cross Flow–Typically, this is a liquid-to-air heat exchanger, such as a duct coil or a convector. If the fluid tubes have fins on them to smooth out the air flow and cut down on the friction loss experienced by the air (the most common design), both the liquid and air flows are said to be "unmixed." The output differs from one in which the air is made turbulent, or "mixed"; and the appropriate chart should be used.

Immersion–Immersion tanks or coils can take innumerable forms, a building being one. As a heat exchanger per se, an immersion coil might be a bundle of tubing inserted into the end of a tank of water; or, as used by the author, a large coil of finned or unfinned copper

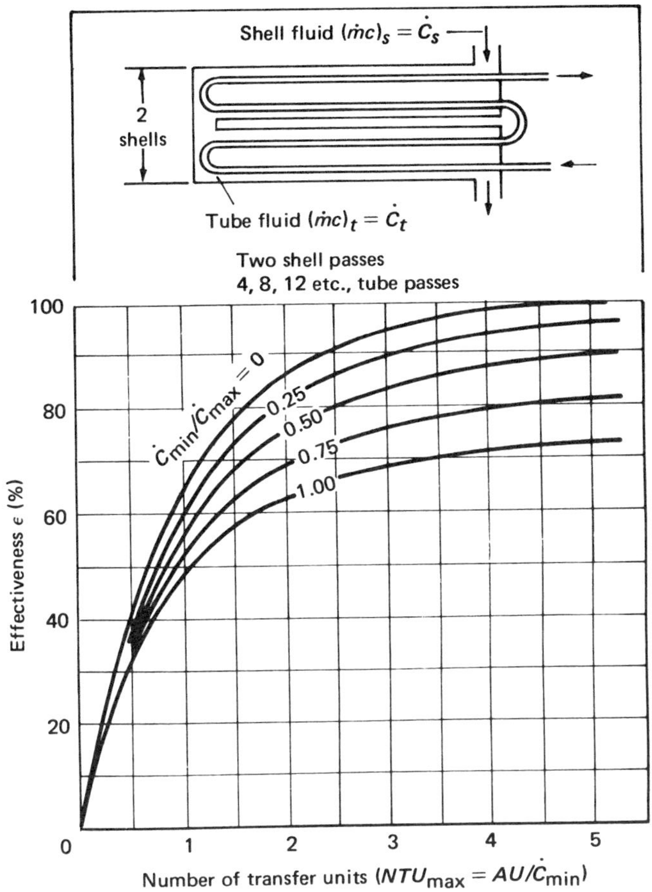

Fig. 5.6 Effectiveness of heat exchangers with two shell passes and four tube passes.

tubing suspended in the solar storage tank. Any one of the charts can be used, following the curve for $\dot{C}_{min}/\dot{C}_{max}$ of zero (since the heat capacity rate of the tank itself is essentially infinite).

In theory, in an infinitely long counterflow heat exchanger with both liquids having the same heat capacity rate, the temperatures of the two liquids could be exchanged, as diagrammed in Fig. 5.7. Such an imaginary device exhibits an effectiveness of 100% (it is of course imaginary, since no heat can flow anywhere in the exchanger because of the identity of temperatures from one end to the other). Suppose, however, that one fluid had a lower heat capacity rate (Fig. 5.8). The maximum possible heat transfer rate between the fluids is in this case

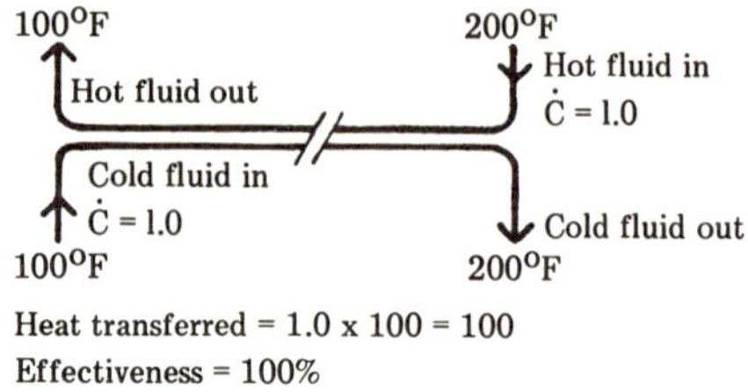

Fig. 5.7 An ideal counterflow heat exchanger with identical fluid heat capacity rates.

limited by the ability of the fluid with the lower rate to pick up or give off heat. In the ideal example shown, the fluid with the lower rate still exchanged inlet temperatures with the other fluid; but the latter did not. The effectiveness remains at 100% by definition, since the device has transferred all the heat possible to transfer, without adding a furnace or refrigerator to the system.

If the difference in flow rates is carried to extremes, we get on the one hand an immersion coil (when the high rate fluid does not change temperature at all); or no heat transfer (when the low rate fluid ceases to flow).

In an ideal parallel-flow design (Fig. 5.9), when the two rates are equal, both fluids exit at the same temperature, halfway between their entering temperatures, exhibiting an effectiveness of 50%. (See Fig. 5.1, where the $\dot{C}_{min}/\dot{C}_{max}$ curve of 1.00 rapidly approaches a 50% effectiveness as an asymptote.) As the rates diverge, the exit temperatures approach the entering temperature on the high rate side, and the device behaves identically to a counterflow design in the limiting case where $\dot{C}_{min}/\dot{C}_{max} = 0$.

The author's preferred way of expressing effectiveness is:

$$\epsilon = \frac{\begin{pmatrix}\text{Change in temperature experienced by the fluid} \\ \text{with the lower heat capacity rate}\end{pmatrix}}{\text{Difference between the entering temperatures of the two fluids}}$$

There are lots of ways to write the basic expression.

A practical example would help de-mystify the problem. Let us suppose that we are unable to avoid placing a heat exchanger between the collector and the store. How large a heat exchanger, and of what design, will maximize system cost-effectiveness? The detailed calculations of cost-effectiveness will appear later. For now, the first step is to decide on flow rates. As noted in Chapter 4, the flow rates through the collectors tend to optimize at high levels. The heat capacity rate on the store side of the collector/store heat exchanger must be *at least* as high as that on the collector side, or little use will come from the increased

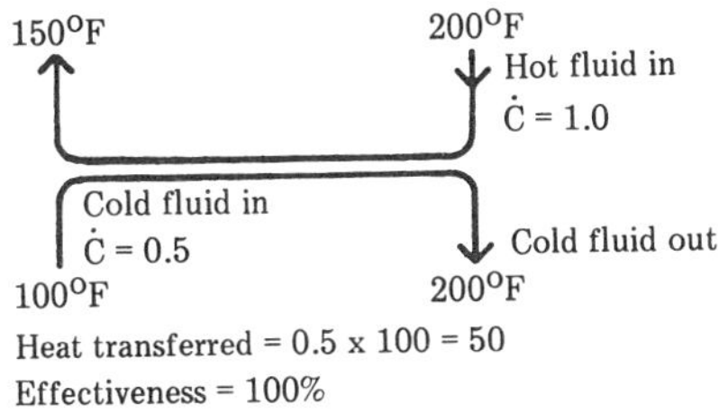

Fig. 5.8 An ideal counterflow heat exchanger with differing fluid heat capacity rates.

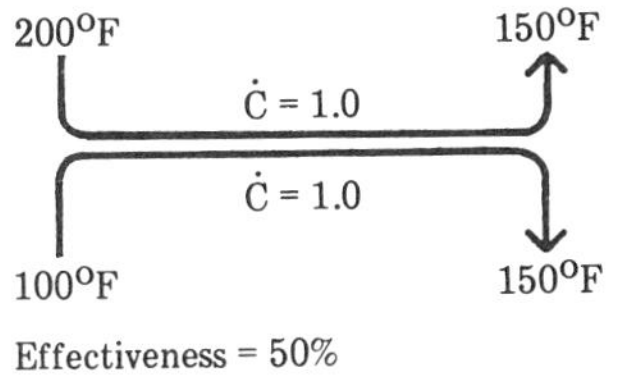

Fig. 5.9 An ideal parallel-flow heat exchanger with identical fluid heat capacity rates.

collector heat-capacity rate. As mentioned earlier, if the rates are equal, a counterflow heat exchanger is the best choice for obtaining good effectiveness. However, an effective counterflow heat exchanger for the typically low flow rates used in solar systems would be prohibitively expensive. Given that we must now use a design with several passes, it immediately becomes apparent by looking at Figs. 5.5 and 5.6 that we must either buy an expensive 2-4 multi-pass exchanger, or a simpler 2- or 4-tube pass design, with a high flow rate on the store side. In fact, since some of the pump energy enters the fluid stream, and the rest may dissipate into a useful space, it might be cost-effective to make the store side flow rate two or three times that of the collector, thus increasing effectiveness without at the same time adding too much parasitic pumping energy cost. It is evident that once a maximum practical flow rate is reached on the store side, the collector flow rate can be optimized to balance the improved effectiveness in the collector (high rate) against improved effectiveness in the heat exchanger (lower rate).

A convenient way to boycott this whole problem is to use an immersion coil. If you first cost-optimize a packaged heat exchanger, and then check how large an immersion coil you can buy for the same price, you will almost always find that the latter has better effectiveness. In the bargain, it avoids the extra pump and pumping energy and is maintenance-free. Its excellent cost-effectiveness is derived from the fact that you already have half the heat exchanger in place, that is, the store itself.

The tricky part of designing with heat exchangers is finding the AU. Many manufacturers do not provide this information, preferring to stick with an older method of sizing which requires that both flow rates and three of the four fluid temperatures be known in advance. If the *AU* is available, it will be in the form of NTU's. If this information is not available, merely ask for the four temperatures and the two flow rates at some condition close to that of the design, and from this data derive the effectiveness directly, using the formula mentioned above. (NTU's can be derived by using Figs. 5.1 through 5.6.)

Immersion coils are more difficult, and the author has been unable to find useful data on the types he recommends—namely, designs in which coils are separated one from the other as far as possible. Each section of tubing sets up a small convective loop in the bath; if the tubes are close together, one loop interferes with the other, reducing the *AU* of the coil. Likewise, in domestic water preheating systems, the water in the coil itself will flow at unpredictable rates. If reduced to a trickle, the water will emerge close to that of the bath, because it has sat in the bath for so long; if moved along brisky, it will cut down the interior film coefficient, and thereby improve heat transfer. If the temperature difference is large, the *AU* will be better than if it is small. As a result, no fixed *AU* can be established for such a coil.

We set out to define a means to express heat exchanger performance in the same terms as a building heat loss rate. Have we achieved this goal? Not quite. If we multiply the building heat loss rate coefficient (*UA*) by the temperature difference between inside and out, we get the actual heat loss rate. If we multiply the heat exchanger *AU* by the average temperature difference between the two fluids, we will get the same answer. However, as in the case of the collector, we only know the temperatures of the fluids going into the heat exchanger. To figure the rate of heat transfer, we put all the elements together as follows:

Rate of heat transfer through heat exchanger =
$$C_{\min} \times \text{Effectiveness} \times \text{Entering fluid temperature difference}$$

(This quantity also equals the product of the temperature difference and the heat capacity rate on either side of the exchanger.)

For example, if a duct coil has a heat capacity rate of 100 Btu/hr × °F and an effectiveness of 75%, the rate of heat transfer would be 750 Btu/hr × °F. It is this value by which all heat exchangers are described in the author's Sun-Pulse II simulation.

GUIDELINES FOR DESIGNERS

1. If a packaged heat exchanger must be used to transfer heat from a storage tank to collectors or a load loop, make the flow rate on the store side as high as practical, to increase the effectiveness. As a rule of

thumb, the collector flow rate will be optimum if it is ½ or ⅓ times the store side rate, and above 0.03 GPM/ft^2 collector.

2. Wherever practical, very large immersion coils, with the tubing spread evenly throughout the tank, almost always exhibit substantially higher effectiveness at lower cost.

3. In using immersion coils for preheating domestic water, a key parameter in the design is the friction drop added to the system by the coil. Make this as large as you can without causing added cost to the rest of the system. The resulting longer runs of smaller diameter tubing will result in a larger area of heat transfer surface.

4. Immersion coils using finned or deformed tubing, which has a high surface area per unit of length, may be cost-effective relative to those using ordinary copper tubing.

5. While no good tables of performance are known to the author, two designs have proved to be cost-effective and to deliver high effectiveness:

For a house: 180 linear feet of 0.5″ I.D. tubing. This reduced the maximum flow rate to about 4 GPM, which may be a nuisance, but which improves heat transfer. The tubing was distributed around the perimeter of a 2000 gallon tank.

For a 90 unit apartment building: Four parallel loops, 240 linear feet each, of 1.5″ I.D. tubing, for a 3″ service main. Effectivenesses above 85% have been recorded by preliminary measurements. The tubing was distributed in horizontal helices in a 10,000 gallon tank.

6. In choosing a collector/store heat exchanger, perform a crude calculation to establish the worth of a degree drop in collector operating temperature (neglect pumping energy, which proves to be a trivial part of the problem). Guess at the number of hours the collector will operate during a year. Guess at the average operating conditions for the heat exchanger. Using the current value of energy, along with your estimate of overall system payback period, establish the investment at that payback for each Btu expected yield. For instance, if oil is being used as backup, at \$10.00 per 10^6 Btu's, and if a 15 year payback is expected, each Btu saved is worth $(\$10.00 \times 15)/10^6 = .015¢$. Now, if the collector is to run about 500 hours per year, and if it has a loss rate of 1.0 ($F_R U_L$), each degree drop in operating temperature is worth $500 \times 1.0 \times .015¢$ or 7.5¢ per square foot of collector. If the array has 400 ft^2 of collector, the 1°F drop is worth $400 \times 7.5¢$ or \$30.00 of investment. Now, merely check the performance of several possible heat exchangers, and stop when the next investment in \$30.00 (including installation) yields less than a 1°F drop in the fluid on the collector side. For instance, if a heat exchanger operating under the standard conditions you have defined has the collector fluid entering at 100°F and leaving at 90°F, it is worth \$300.00

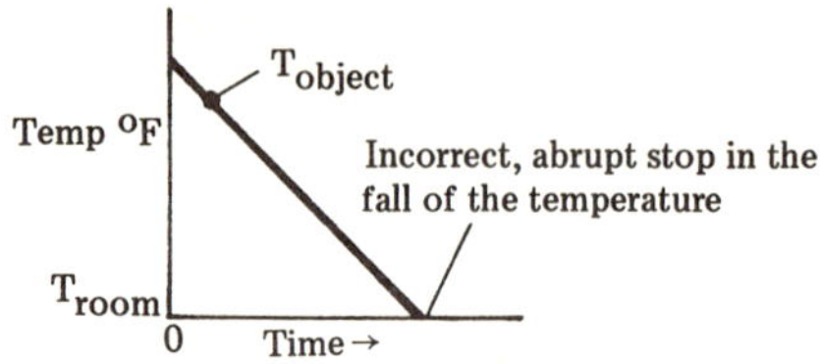

Fig. 5.10 An incorrect guess of the cooling rate of a warm object in a cooler environment. The guess is based upon a steady-state analogy.

more than one where the collector fluid enters at 110°F and leaves at 100°F (10°F difference × \$30.00 per °F).

7. If you are using the storage tank to preheat domestic water, some regulations insist that you arrange the piping so that leaking antifreeze does not come in contact with the domestic water piping. This severe rule has been promulgated to protect users from poisoning in the unlikely event that both sets of pipe leak, and that the initially nontoxic antifreeze (if such is used) is later replaced by toxic antifreeze. To accomplish this legerdemain, you either need a double-wall coil, in which the outer wall drains out of the tank; or you need to place one of the heat exchangers outside the water in the tank. Many designs for water heating tanks include a wraparound heat exchanger on the outside of the tank, which functions reasonably well, although not as effectively as an immersion coil. Avoiding this problem is one of the primary advantages of a drain-back system which does not use antifreeze.

EXPONENTIAL DECAY AND SYSTEM TIME CONSTANTS

Abundant and cheap supplies of energy have led us to expect our buildings to be kept at constant temperatures. Designers of heating systems generally consider steady-state conditions, because they merely have to be sure the heating plant is large enough to satisfy the demand under the worst condition (or under some large percentage of conditions). "Failure" of the system, aside from total shutdown due to malfunction, consists of falling behind slightly for a few hours during the coldest night—not a very disturbing prospect.

As we explore more natural ways to heat our buildings, such as passive heating and cooling, we find that the essence of the problem involves *transient* conditions. Interior temperatures might not stay even; heating supplies might depend upon the condition of the building itself rather than the condition of a separate power plant. If we study the *time history* of many kinds of systems including solar heated systems, we come again and again upon an important question: how rapidly does

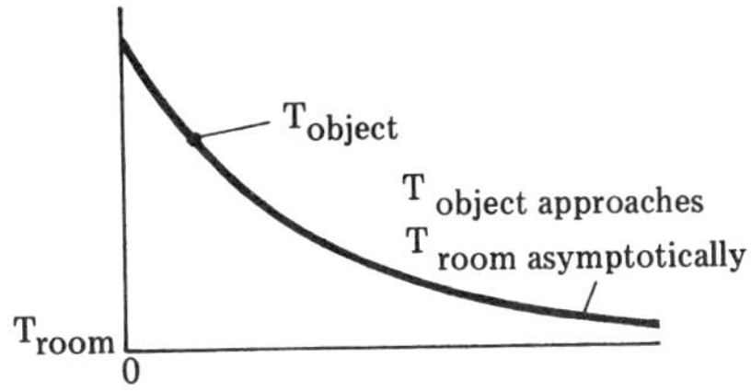

Fig. 5.11 The correct guess of an object's cooling rate over time.

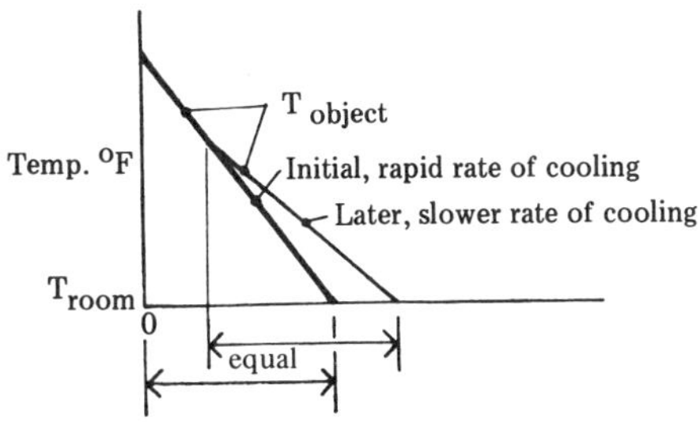

Fig. 5.12 The cooling rate slows down as the object gets cooler.

something cool off or heat up? The question arises when we examine the discharging of a solar storage tank to a house, or a house itself, if unheated, to the outside environment. Being a body immersed in air, a house resembles more familiar objects which are easier to study. One good example is a cup of coffee.

Suppose you have heated a cup up to the temperature of hot coffee, then filled it with coffee and left it to cool on the kitchen table. We are certain it will cool off, but uncertain about how fast. A simple guess as to the graph of the cooling process over time is shown in Fig. 5.11. It fits our predilection for predicting future change at the same rate as that of the present, the same predilection which leads us to misunderstand exponential growth. The graph just has to be wrong, because we remember that the coffee cools quickly to a warm state, stays warm long, and doesn't abruptly stop cooling. So we might try a curved graph like Fig. 5.11. Such a graph proves to be accurate and is an example of a wonderful curve called an "exponential decay" curve. It is mathematically similar to the exponential growth curve made popular in "Limits to Growth," but it has a number of special properties.*

1. The curve is asymptotic to the base line and so mathematically never reaches zero. Because of the statistical nature of heat, it in fact *does* reach a point where you can't measure the difference.

*To an engineer who uses exponentials daily, their properties are second nature. To the nonmathematician, looking back at exponentials years out of math class, their properties are both magic and beautiful.

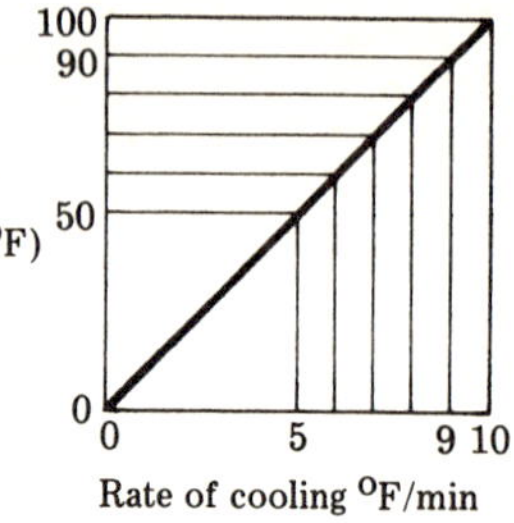

Fig. 5.13 A graph of the cooling rate of an unspecified object in a cooler environment, as a function of the temperature difference. The rate is a linear function of the temperature difference.

2. The curve is based upon an empirical observation which is close to the truth for many kinds of systems, namely that the rate at which the cup of coffee in the example cools is proportional to the difference between its temperature and that of the room.

To see how this latter observation determines the nature of the decay curve, observe that the rate of decay determines the slope of the curve. If the coffee is cooling rapidly and were to continue to cool at that rate, it would reach zero in a short time. If it were to do so at a constant slow rate, it would reach zero much later. But in fact it cools at a certain rate for the time period and then, being cooler, cools more slowly during the next time period. This is shown in Fig. 5.12.

Suppose the coffee initially cooled at the rate of 10° a minute. If it started 100° hotter than the room, it would cool to room temperature in 10 minutes. But, after one minute, it has cooled 10°, to a temperature 90° hotter than the room. How fast will it cool now? We said the cooling rate was proportional to the difference in temperature. Before, it cooled at a rate of 10° a minute when it was 100° hotter than the room. It is now 90° hotter than the room and will therefore cool at 9 degrees a minute. (See Fig. 5.13.)

Ask another question: how long will the coffee *now* take to cool to room temperature, at the new rate? Answer: 90/9 or 10 minutes! This result obtains for every temperature and every rate—if the coffee continued to cool at the rate of the moment, anywhere on the curve, it would finish cooling in a constant amount of time. We can, in fact, draw a good approximation of the exponential decay curve by using this constant time period, as in Fig. 5.14.

As with any continuous process, if we choose very small time intervals, infinitesimally small ones, we come up with a smooth curve. The time interval which is constant along the curve is called the "Time Constant." In this example, the time constant is 10 minutes. It is also the ratio between the temperature difference and the rate of temperature change in our proportionality graph: 100° ÷ 10°/minute = 10

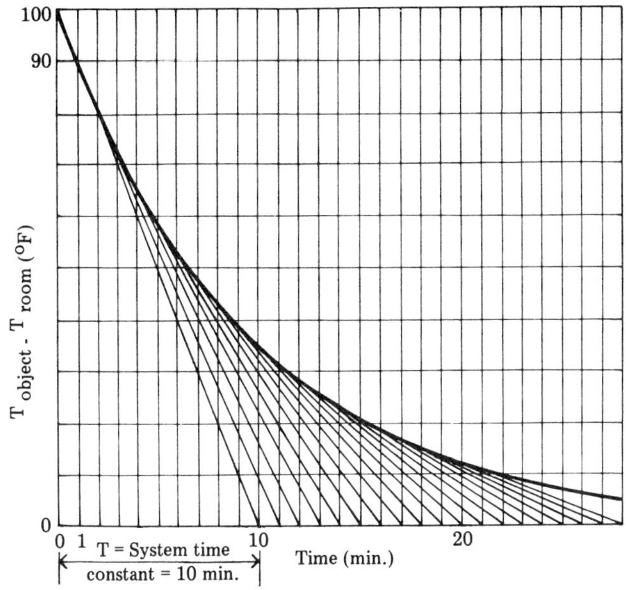

Fig. 5.14 The exponential decay curve is a consequence of the linear relationship between cooling rate and temperature difference graphed in Fig. 5.13.

minutes. In a way, the time constant determines the "steepness" of the curve—a long time constant means a slowly changing system; a short one, a rapidly changing one. Exponential curves are wonderfully liberating mathematical constructions, and the reader is invited to explore them further. We need now to find out what characteristics of the system determine the system time constant.

THE RELATIONSHIP BETWEEN CAPACITANCE AND CONDUCTANCE

Compare two systems. The first is the one described in the previous section—a heated cup filled with coffee heated to the same temperature and left to cool on the kitchen table. The second is the same, but without the coffee. Which will cool more rapidly? Intuitively, we feel that the cup without the coffee in it will cool more rapidly because there is less of it to cool off.

Now add a third cup, heated like the other two to the same temperature as the coffee, then filled with coffee and wrapped in insulation. We are sure that this cup will cool more *slowly* than the uninsulated cup of coffee, and that the more insulation, the slower the cup will cool (Fig. 5.15).

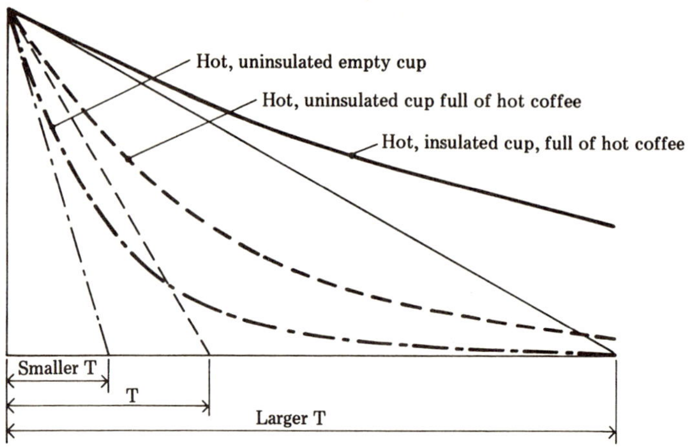

Fig. 5.15 The system time constant is proportional to its capacitance and inversely proportional to its UA coefficient (or, if you like, proportional to its resistance to heat loss).

Generalizing, the time constant depends upon the amount of thermal energy in the system, and upon how fast heat leaks out. The energy in turn depends upon the *mass*, and upon the *specific heat* of the material. In the example above, the coffee cup had less mass than the coffee, and it had a much lower specific heat, the effects combining to reduce its time constant. Heat leaks out very slowly from the insulated cup producing a long time constant. From previous discussion, we know that the rate at which heat leaks out of a membrane is measured by the membrane's *U*-factor.

Mathematically, the relationships can be expressed:

$$\text{Time Constant} = \frac{U\text{-factor} \times \text{area}}{\text{mass} \times \text{specific heat}}$$

Applying those ideas to some examples within solar-heated systems will help to nail down the ideas. If an uninsulated masonry house is heated up, allowed to cool for a while, and then its time constant is measured, that value might very well equal the time constant of a lightweight, well-insulated house put through the same thermal conditions. But when the time comes to reheat the two houses, the lightweight house requires much less *energy* than the masonry one. In slowing down the rate of thermal decay, the masonry house in a sense threw all its energy troops into the breach, while the insulated house built a solid defense and used its troops sparingly.

When we measure the thermal behavior of a storage tank, we describe the *amount of heat* the tank can absorb for every degree its temperature increases: Btu/°F. When we measure a heat exchanger, which is the device which *charges or discharges* the store with heat, we describe the *rate of change* of energy per degree, that is, the power transferred through the heat exchanger per degree: Btu/hr × °F. If we divide the storage measure by the heat exchanger measure, Btu and °F cancel out, and we get only hours, which is a measure of *time*. This is in fact the time *constant* of the subsystem consisting of the store and the heat exchanger. We know that as the store discharges through the heat exchanger its temperature decays exponentially, and that the energy discharged from the system in a given unit of time (power loss) also decays exponentially. The relationship of this exponential decay, or more exactly its time constant, to the rest of the thermal transactions taking place in a building, can to a large measure determine how efficient a solar heating system is.

In practice, since both the store and heat exchanger influence the system time constant, you cannot successfully deal with one or the other in isolation. They must *together* be optimized for cost and size, because together they determine the system time constant. Not understanding this principle has led most system designers to undersize their stores, because they did not see the importance of enlarging the load heat exchanger and allowing it to operate at lower temperatures. This in turn would lead to larger, lower temperature stores and a consequent improvement in overall system efficiency.

In the discussion so far, we have talked about simple systems undergoing a regular, steady decay in temperature. When we look at a house, we find a much more complex situation. Rather than a simple body of fluid surrounded by a membrane, the house consists of a series of plates distributed around the air-filled volumes. In old or leaky houses, some outside air penetrates the wall spaces, making the situation more complex. But even in a very tight, well-insulated house, there is a fundamental issue which makes the determination of a time constant for an entire house difficult or impossible, and which bears upon the processes occurring in passive heating.

Under normal circumstances, a house is kept at a constant temperature by adding heat at approximately the rate the house loses heat. As discussed later, the inner layer of the outside walls, and of course the glass, are at temperatures intermediate between inside and outside; while the wallboard, wood, and masonry within the house are basically at room temperature. If the heat is turned off, the room air quickly falls to a lower temperature, until the heat stored in the interior walls and in the interior layer of the exterior wall begin to give up enough heat to make up for the air's loss to the outside through leakage, conduction through glass, etc. This is obviously a complex process taking place over

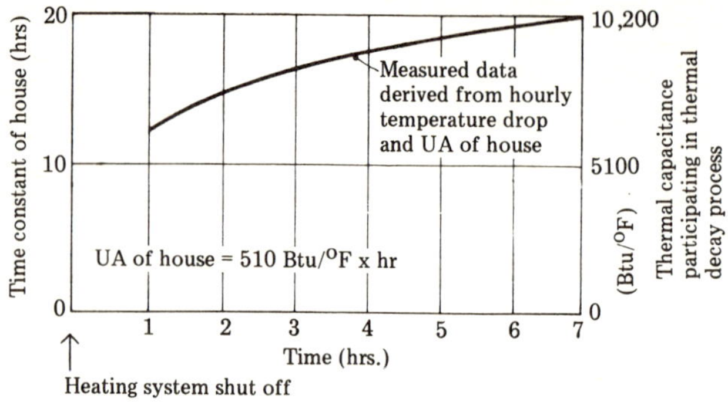

Fig. 5.16 Graph of the thermal decay "time constant" (it is not a constant at first) of the Acorn Test House as a function of time, after shutting off the heating system. Because the heat loss rate is fixed, the rate of decay is linearly related to any change in the capacitance. As the house cools, more and more interior mass is involved in the decay process, and the building approaches a true steady-state exponential decay with a true time constant. The data was generated by averaging the hourly temperature readings; the first hour is therefore not shown. Theoretically, the graph begins at the origin. It is important to note that in a passive building, most of the mass may be *above* room temperature in the first place, a condition equivalent to a normal house after several hours of "coasting" without heat.

time, and the system will not "settle down" until many internal transactions have been completed. The graph of the interior temperature in the Acorn Test House (Fig. 5.16) shows the "settling down" process during an actual test, in which the heating was turned off for several hours (the fan was left on, to prevent stratification of the air).

One interesting question arises: what happens at the instant the heat goes off, even during a normal cycling of the furnace? Since the air had been held at an artificially high temperature by an outside heat source, it must fall to a lower temperature before the heat stored in the building mass can flow into the air and thus make up the loss to the outdoors. So, viewed superficially, the heat-loss rate at the instant of turn-off is determined only by the mass of the air. Examining the processes within the house more closely, the various surfaces are at a range of temperatures so that the house cannot be considered as a single object but rather as a collection of surfaces all exchanging heat with each other, by radiation, by conduction, and by convection through the air. When the furnace turns off, these processes continue and some participate immediately in keeping the house warm. But, as seen in the Acorn example, the initial rate of temperature decay is indeed much faster than the rate which prevails after the house has settled into a steady-state condition.

This rapid initial rate of temperature change, combined with our expectation that houses are very leaky and suffer rapid drops in air temperature before the stored heat is brought into use, lead us to underestimate the possibilities and problems of a massive, well-insulated house left to "coast" on its own. An example from Massdesign's experience is relevant here. In a very well-insulated building with considerable mass, the thermostat was turned down habitually at night to 55°F. A counter separated a large, triple-glazed wall from the room thermostat, allowing a pool of cold air to collect under the center of the glazed wall. Because it was not extremely cold outside, the building coasted for several hours before calling for heat. During this time, the heating pipes under the windows burst due to the pool of subfreezing air which had collected. Put to a better use than flooding valuable property, the ability of such a building to coast could figure profoundly in the design of heating systems.

6

HEAT LOSS AND INSULATION

THE BUILDING AS A HEAT EXCHANGER

In a solar heated building, heat is exchanged between many system elements:

- The collector loses heat to the environment and to the working fluid.
- The store gains heat from the working fluid, loses heat to *its* environment, and loses heat through the load heat exchanger to the building air.
- The building loses heat to the environment and gains heat from innumerable other sources, including the back-up heater and the solar load heat exchanger.

All these transactions take place by means of some sort of heat exchanger, although not usually of the sort specifically so named. (Some general comments about heat exchangers are found in Chapter 5.) One of the numerous kinds of "heat exchangers" is a warm tank within another fluid—a building, neglecting solar gain, is of this type. While heat exchangers are built to *maximize* the rate of heat transfer across a membrane, a building wall is intended to *minimize* this rate. But the same issues are involved.

Both the building and the environment change temperature over time. The effects of those changes are of great interest (see Chapter 7). To study its behavior as a heat exchanger, it is not necessary to look at such transient behavior, but only at a steady-state condition, one which might persist for a period of time or change only slowly to another state. Assume that the building thermostat is set at 70°F, and that a good ventilation system keeps the air in the house from stratifying. What happens at the house walls?

Every Btu of energy which leaves the house must be replaced. The air is the medium which accomplishes this, as it moves slowly around the room and through a baseboard convector or return air duct, passes over the hot surfaces of a heat exchanger, and returns to the room. All the room air need not be heated by the furnace or convector; some of it can be overheated and mixed with the room air to produce the desired average temperature. Real furnaces and space-heating units do not usually run continuously but cycle on and off, alternatively overheating and overcooling the space. Assume for this analysis that good control has matched the output of the heating system to the loss through the walls.

As the room air approaches the outside walls or glazing, it encounters a layer of still air, more or less "glued" to the inside surface of the outside wall. If the room air is moving slowly, this layer can get quite thick—perhaps a half inch. Of course, it has no well-defined boundary, since the air slows down gradually as it approaches the wall surface. On the outside of the wall, the wind erodes this layer of still air.

This "air film," as it is called, insulates the outside wall to some extent. Aside from the absence of air, air is really our only useful insulation material. The human body insulates itself in part by raising hairs on the surface to increase the depth of still air over the skin—the phenomenon of goose bumps results from the raising of the hair follicles to extend the skin hairs outward. We could in fact insulate a house by wrapping it inside and out with hair or fur (wait until the architectural avant-garde finds out about this!). We don't, simply because such materials are a maintenance problem if left exposed. Rather, we place the fur on the inside of the wall in the form of light material, loosely connected, which traps air (or another gas) in small cells. We call such material insulation, and it ranges from seaweed and old newspapers to shiny films, cellulose, various plastic foams, and fiberglass.

Heat can be transferred by convection, conduction, or radiation, and typically two or three of these mechanisms work together. In the space, room air moves heat around by convection, but a fireplace or other heater in the room works by radiating heat to the surrounding surfaces and people, while only conduction can move heat through an opaque solid. The insulating value of a material or an air film aggregates all these mechanisms into a single coefficient. Under most circumstances, this is an accurate procedure, but exceptions can be found—for instance, a fireplace heating a wall directly can make up the entire heat loss without any contribution by the room air at all.

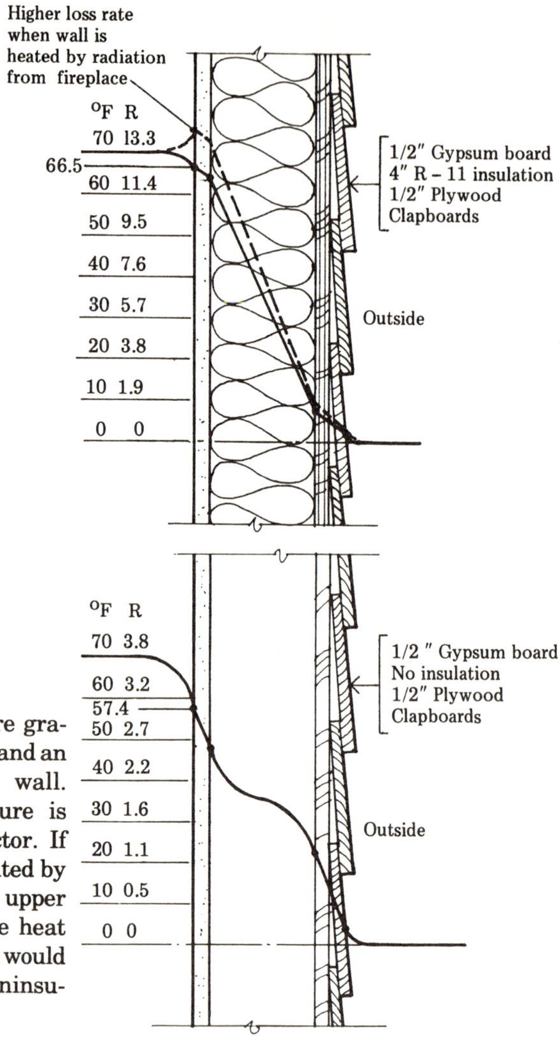

Fig. 6.1 The temperature gradient within an insulated and an uninsulated wood stud wall. Note that the temperature is proportional to the R-factor. If the interior surface is heated by radiation, as shown in the upper diagram (dotted line), the heat loss is increased. The effect would be very noticeable in the uninsulated wall.

It is a crucial fact that many heat exchange processes are "linear." This means that the rate at which heat is transferred is proportional to the temperature difference between the cold side and the hot side. For example, heat flows out of a 70° room twice as fast if it is zero outside as it does if it is 35° outside. Because of this approximately true observation, we can assign a simple coefficient which relates the rate of heat loss with the temperature difference. Such a number, with dimensions of Btu per hour per degree of temperature difference per square foot (Btu/hr × ft^2 × °F) is called the "U" factor. Its inverse, with inverse dimensions, is the "R" factor. It measures resistance (hence the R) while U measures conductivity. Since this is a proportionality constant between heat loss and temperature, a graph of the U-factors in a wall can be interpreted either as a graph of temperatures or one of heat loss. Let us see how this works.

The U factor for an assembly of materials is found by counting up the R factors, and then taking the inverse of that sum. For instance, in the upper part of Fig. 6.1, the R's are as follows:

air film	0.68	
gypsum board	0.5	
insulation	11.0	
exterior wall	1.0	
exterior film at 15 mph	0.17	
	13.35	U = 0.075

It follows from the linear relationship between temperature and heat loss that the temperatures at any point within the assembly are proportional to the R factors on either side. Thus, if you penetrate to R 6.67, you are halfway through the temperature drop. The diagrams in Fig. 6.1 show the temperatures plotted in this way.

Compare an "insulated" and "uninsulated" (i.e., intentionally insulated and unintentionally, partially, insulated) wall (Fig. 6.1). Notice that the interior film coefficient in the uninsulated example constitutes a substantial percentage (18 percent) of the resistance, whereas in the insulated example it accounts for only 5 percent. By proportion, the temperature drops only 5 percent through the film in the insulated wall, from 70° to 0.95 × 70° = 66.5°; whereas in the uninsulated wall the temperature drops from 70° to 0.82 × 70° = 57.4°. The interior surface of the uninsulated wall will be 18/5 cooler, with the air in the room losing heat to that surface 18/5 faster.

In general, because heat is not accumulating anywhere as it passes through the wall assembly, any section of the wall must have a temperature drop which is proportional to its R factor. For instance, the drop through the insulation, which accounts for 11/13.35 of the heat loss in the better insulated wall, must be 11/13.35 of 70° or 57.67°F. This explains why a fireplace can heat a wall by itself. The radiation passes through the transparent air film, raising the inside temperature of the wall up to 70° or even higher. The air in the room "sees" a wall at or above its temperature—if above, the air is heated by the wall. Now, the assembly consists of a wall minus the interior film coefficient, losing heat across a 70° or higher temperature difference, with the losses made up by the radiation from the fireplace. Because the walls are warmer, radiation from the bodies of the occupants to the wall is impeded, thus making the room feel cozy and warm, and perhaps inspiring the occupants to turn down the thermostat. This will not slow down the heat

Fig. 6.2 The "Saskatchewan House." Later versions, without the complex active collector, are nearly 90% solar in a 11,000 degree-day climate. The infiltration rate is reported to be around 1/20th air change per hour, requiring an air-to-air heat exchanger to maintain room air quality. Photo courtesy of J. W. Besant. For a detailed description, see the May 1979 issue of "Solar Age" magazine.

loss; in fact, since the interior surface of the wall is now warmer than before, heat loss actually increases, with the fireplace supplying the new energy (Fig. 6.1).

A similar effect occurs on the outside when the sun is shining. A rough-surfaced, dark-colored wall can be heated by the sun to relatively high temperatures, even on cold, windy days. The wall acts like a badly designed solar collector. By raising its outside temperature, the sun prevents heat loss, or actually produces a slight gain to the interior. In the example of the insulated wall, the loss rate was 70/13.35 = 5½ Btu/hr/°F/ft^2. If the sun could raise the exterior temperature to 70°, this energy would be saved. For a 300-square-foot house wall heated by the sun for six hours, the saving would be nearly 10,000 Btu, worth 28 cents at 10 cents per kilowatt hour.

To measure the overall skin losses through the exterior walls of a building, it is convenient to arrange the figures to yield a loss rate in Btu's per hour per degree Fahrenheit. From this, one can either determine the maximum loss rate in Btu's per hour (given the maximum temperature difference), or can estimate the losses over a time period by establishing the number of degree-hours or degree-days of loss. Also, in the basic form, other losses can easily be added to the skin losses.

Besides losses through the skin by a combination of conduction, convection and radiation (collectively called the "conductive heat loss") described above, important losses occur because of intentional or unintentional ventilation of the building. There are two ways to estimate ventilation losses for simple buildings: either by adding up the known cracks around windows and doors and in corners, dividing by two (since air must both enter and leave), and assigning a heat loss value to the cracks; or by figuring the number of total air changes which occur in the room space. The air change in question is not the internal ventilation air change, which keeps the air from becoming stagnant, but the quantity of outside air entering and leaving the building. This heat loss is simply the number of air changes per hour times the amount of air times the heat content of a unit of air, in Btu/hr × °F (the same units as the conductive skin losses). Typical values range from 2 or more for a leaky old house down to preposterously low values like 0.05 for super-insulated houses such as the "Saskatchewan" house shown in Fig. 6.2.

Unintentional ventilation losses, usually called "infiltration losses," are a function not only of temperature, as noted above, but also of wind pressure, which varies with the square of the wind velocity. Shielding a house from the wind helps hold down the pressure, although the suction can induce substantial losses, as can the "chimney" or "stack" effect in tall structures, caused by internal temperature and pressure differences.

Important losses also occur between the building and the ground. Since the ground stays at a relatively constant temperature, these losses are *not* linear with outdoor temperature. They are usually relatively small and can be approximated by an average winter-time value.

Another nonlinearity in the simple model presented above is the effect of wind on the conductive losses through glass. Since ordinary double-glazing is poorly insulated, with an R value of about 1.6 or 1.7 compared with an ordinary insulated stud wall with an R of 13 or so, the influence of the exterior film coefficient is correspondingly larger. By convention, that coefficient is taken at a wind speed of 15 mph, which is a high average for many parts of the country. In still air the insulating value can increase by 25 percent because of the thicker air film.

A CLOSER LOOK AT HEAT LOSS AND THE EFFECTS OF OCCUPANCY

The appealing and measurable picture of building heat loss presented above is sufficiently complex to need a book of its own. What actually goes on in a residence, and to some extent in other buildings, is mind boggling by comparison.

- In real buildings, the air stratifies in unpredictable ways.
- Building maintenance determines the value over time of weather-stripping, sealant, etc.
- Buildings move, materials expand and contract, surfaces weather, metal corrodes, flashing deteriorates.
- Various insects, birds, and rodents do their best to destroy the building or convert it to their needs.
- The occupants are busy moving around, cooking, turning lights on and off, opening and closing windows and doors, changing thermostat settings, going on vacation, etc.
- A virtual arsenal of appliances is being used or run on their own, at various times of day and at different times of the year.
- Moisture is evaporated from a variety of sources, sometimes deliberately, adding a load for the heating system.
- The heating plant, often located in the basement, loses heat through the furnace or boiler walls and through the piping or ductwork, to the basement, which may or may not be insulated well.
- If present, the solar storage tank loses heat either to the living space or the basement, or to the ground or outside air.

Before lumping these factors into neat packages, let us look at them in more detail.

Stratification

Air in buildings stratifies (forms layers), since hot air is less dense than cool air. In larger buildings with more sophisticated heating systems, measures are taken to mix the air carefully, unless the designer uses high returns to withdraw heat from the lights before it increases the cooling load to the space. In residences, more primitive ventilation systems suffice, and layering is common. Designers increasingly install high returns (along with, rather than instead of, low returns) to reduce this stratification. Oldfashioned, slow-moving, "tropical" ceiling fans are being used, especially by owners of wood stoves, to drive warm air back down to people height.

What stratification exists affects the building heat loss in several ways:

- Codes based on ASHRAE 90-75 require that ceilings be better insulated than walls, to account for the stratified warm air at the ceiling as well as to inhibit summertime solar gain.
- While the average space temperature is always used to measure heat loss, in fact the upper walls lose more and the lower walls less

heat than average. Suppose for some reason a designer specified little glass at floor level and a great deal in clerestories (a possible strategy in a dense urban setting). The higher temperature at the clerestories would cause a greater overall loss than that calculated using averages.

- When measuring the ability of a solar storage tank to discharge through a duct coil, the temperature difference between the incoming water and the incoming air determines the coil's "effectiveness" (see Chapter 5). In fact, the incoming air is return air, which is inevitably at a lower temperature than the average, improving the coil's effectiveness.
- If a house is designed with rooms to be left at lower temperatures, they ought to be downstairs rather than upstairs. Put another way, if partly heated rooms are located upstairs, they should be insulated and weatherstripped, or else they will be warmer than anticipated, thus raising the actual heat loss above the design goal.

Ventilation

When unwanted ventilation is reduced to a minimum, a new problem arises: how much ventilation do you need? Five years ago a house needed at least one to two air changes per hour. Today estimates go as low as 1/4 air change per hour. Surely we could not have cut our need for ventilation by a factor of four in as many years!

People, thank heaven, are not all alike. Some hate the smell of cabbage, yet cook it. Others don't mind the smell of cabbage, but don't cook it. Some smoke and don't mind the odor clinging to everything. Others don't, or used to, and can't stand the slightest suggestion of tobacco odor. Some are casual about bathroom matters and prefer sulfuric odors to the introduction of piney scents; others are distressed by the slightest suggestion of the presence of plumbing. Finally, while some people can spend an entire winter with windows sealed, others can't sleep if new air is not entering the bedroom.

It is tempting to jump in and design a giant computer model to analyze all these variables, or at least to perform studies to find out how the mythical average family actually behaves. While such data are sometimes useful (and may get you a government grant), it is important to remember that we are not looking at an uncertainty in *measurement* but an uncertainty in *reality*. Suppose, for instance, that an owner with the best will in the world promises that he will adopt every energy conservation measure possible, and that on the basis of this the designer promises that the solar system plus solar window gain plus a little wood stove will take care of the heat loss. The owner then gets sick, or sells the house, or forgets what to do—and there is inadequate heat available. Was the designer wrong? What is the proper house-design heat loss?

This dilemma can be solved by establishing one set of parameters to represent the effects of occupany, and another set to represent the inevitable occurrences which follow from the climate and the construction of the building. A good design procedure will allow the designer to change the occupancy parameters in order to exercise a proposed system or design over a range of occupants, just as the building itself might be exercised in reality. It is important to note that the variability of occupancy is much more pronounced in single-family residences, where the occupant is the lord of his own castle, than in corporate or institutional buildings, where a central management can and does determine who does what when.

With all this diversity, it is no wonder that we have a variety of "standards" to choose from. In a time of abundant energy, we could point to the worst condition as standard. Now, we look to the relatively insensitive energy miser as a model.

Another way to look at this issue is to examine the household appliances which lose energy through ventilation, and to assign a value to these losses. How often is a kitchen or bathroom exhaust fan used? How much of the time is the front door open? Or the bedroom windows? How do these quantities change during the year? While endless studies have tried to find out these things on the average, little has been done to correlate any of these data with real people so that the designer can predict with any accuracy what will happen in a particular building.

Heat wheels and other ventilation air-scavenging devices have been used for some time in nonresidential buildings. As we move toward super-insulated houses with very low ventilation rates, the use of small air-to-air heat exchangers in houses may become essential, especially in light of recent findings that radioactive gas can accumulate in closed spaces over time. Loss rates can be kept very low, with excellent ventilation being provided for crucial areas.

The Use of Lighting

Many people forget to turn out lights in unoccupied spaces. An average electrical consumption figure would include many such people and would not reflect the average if electricity were two or three times as costly as it is now. Lighting also is not as efficient as it could be within our high standards of aesthetic acceptability. If a large figure is used for the *heating* effect of lighting, it might prove to be inaccurate in the future as occupants learn to conserve lighting energy.

The Use of Appliances

While microwave ovens and garbage compactors are gaining acceptance, the post-World-War-II explosion in household-appliance energy use has tapered off and begun to fall with the introduction of more efficient appliances (air conditioners are *not* an appliance in the sense meant here). While changes in the design of appliances will have a gradual influence on the waste heat available, the occupant can abruptly decide to wash clothes in cold water, handwash the dishes or turn off the automatic defrost cycle on his refrigerator. Again, the heating effect will be reduced, with solar or the heating system taking up the slack.

Cooking habits are not likely to change dramatically, but they have a definite ethnic and age-related dimension which is poorly studied. If a building has an identifiable occupancy, such as elderly housing in a community where cooking is an important social activity, the designer can expect a substantial contribution from cooking heat.

Humidification

When humidification is a requirement for comfort in the heating season, it cannot be considered an adjustable load. However, earlier studies indicating that raising the humidity allows lowering of room temperature appear to have been at least partially mistaken, so that the minimum level of humidity in a residence is largely a matter of taste. Some doctors think that low humidity contributes to ill health, and most doctors will prescribe humidification of a patient's bedroom to relieve certain symptoms. Some articles of furniture, such as harpsichords, antique furniture, and art works, need controlled humidity conditions. Besides being hard to measure, humidity is not stored very long. Both factors make it very hard to define the heat loss in a residence which results from humidification. It is best handled by derating the heat gain from occupancy, on the reasonable assumption that cooking will supply most of the humidity.

Occupancy Patterns

People's going in and out has a major impact upon the house heat loss, largely through potential reductions in thermostat setting, but partly due to the sensible and latent heat gain from the occupants themselves. In vacation houses, the pattern of occupancy assumes major importance and can by itself determine the feasibility of solar heating. Turning the thermostat way down during the day is undoubtedly an increasing pattern, certainly in working families with no one at home, but even in some cases where "empty nesters"—couples with grown children—occupy only part of the house and leave the rest cold.

Zoning the heating system is itself a reasonable way to save energy and will become an increasingly common pattern.

The basic setting of the thermostat during heating hours shows signs of a steady reduction, as Americans begin to believe they have an energy problem.

In calculating the useful gain from occupancy or south glazing in any structure, the *location* of the gain must be known. Such calculations are commonly made for commercial structures, but seldom for residences. Consider a small house or apartment with a partially or fully enclosed kitchen. Not only the range, the refrigerator, and the dishwasher, but sometimes the clothes washer and dryer are located therein. If the room also faces south, overheating will be common all year round. Unless measures are taken to transfer the excess heat into storage (*not* located in the kitchen itself) or into other spaces, much of the building's occupancy heat gain and some of the south window gain will be wasted, even if the rest of the house is cold. In setting these parameters for a computer program or hand calculation, either the excess energy must be discounted, or the building heat loss adjusted to reflect the overheating in the kitchen.

For larger structures, a careful analysis of internal gain is required. An increasing number of office buildings are using energy scavenging techniques, with storage, in some cases eliminating the need for a separate heating plant. Enclosed pools have another set of problems related to the massive "humidifier" in the space, and the need to dispose of excess vapor. Industries with internal process heating may need only measures to recover some of the lost heat. A wide variety of building types, each with its own characteristics, are well-covered elsewhere in the literature, and cannot be dealt with here. In general, larger buildings have cooling rather than heating problems.

Spaces in Contact with the Ground

Conduction losses through the ground vary independently from that of the rest of the house. First, consider the unheated basement. In order to discover the heat loss from the house into the basement, a heat balance must be run. The heat gains are from the living space above, and from whatever heating appliances are present. The losses occur normally, through windows and walls exposed above grade; through walls in the transition from grade to normal ground temperature; and through walls and floors to ground at normal temperature. The U-factor of the ground varies with the type of soil geographically and sometimes at different points in the foundation of a house. It also is highly sensitive to the presence of groundwater and to the effectiveness of various measures taken to control it. As shown in the charts in Fig. 6.3, the presence of insulation in various places throughout the skin of the basement influences the variation in heat loss as a function of outdoor temperature.

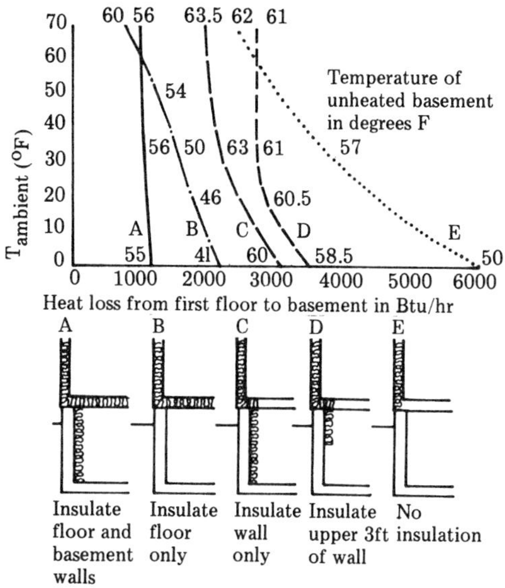

Fig. 6.3 Chart of the heat loss between the first floor and basement, and the resulting basement temperature, for various outside temperatures and various insulation schemes. The house used for calculation had a conventional heating plant losing about 10% of its energy to the basement of a 26′ × 44′ home. The ground was assumed to be at 50°F, with an R-factor of 10. No attempt was made to model earth storage.

Such a chart shows that despite the complexity of the situation it is possible to come reasonably close to the seasonal basement temperature, which tends to remain relatively constant. More study is needed.

Crawl spaces usually differ from basements through the absence of heating appliances. However, crawl spaces should be ventilated except in winter, and many owners fail to close the vents during winter. If that is likely to be the case, the crawl space should be treated like the outdoors, with heavy insulation in the floor. Piping naturally would run above the insulation, or be insulated and possibly traced with electric heating wire. (Tracing should be thermostatically controlled by a sensor near the wire and should be thoroughly covered by insulation. Exposed tracing is a classic energy waster.)

Slabs on grade present the same problem as basement floors with regard to determining the U-factor. In addition, the edges of the slab, if uninsulated, conduct heat very effectively to the outdoors. In new buildings, positive insulation should continue from the center of the exterior wall down past the slab to one or two feet below grade, depending upon the severity of the winter temperatures (Fig. 6.4). While the insulation can be run under the slab instead of down the foundation wall, if water is

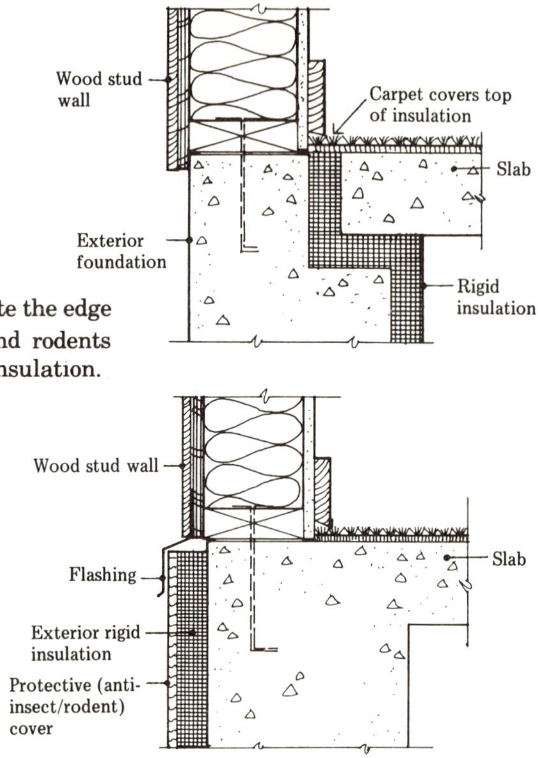

Fig. 6.4 Two ways to insulate the edge of a slab on grade. Insects and rodents appear to love exposed foam insulation.

present below the insulation heaving can occur at the slab edge. Some designers now recommend putting the insulation outside the foundation, joining the wall insulation above grade, and including the foundation mass in the thermal inertia of the house. There seems to be little rationale for this practice, unless the entire foundation and slab is insulated from the ground. If the ground itself is used for thermal storage, why bother adding the relatively modest additional storage in the foundation walls? But in any case, exterior foundation insulation should be protected from moisture, rodents, and carpenter ants, which appear to think that polystyrene is the finest housing material ever invented.

Energy Interactions

Certain energy conservation strategies have attendant increases in energy consumption elsewhere in the system. For instance, reducing lighting saves electricity and reduces the cooling load in summer but increases the load in the winter, since the heat dumped into the space by excess lighting must now be made up by other, presumably cheaper, energy sources. Even if little is saved in winter, the summertime saving in energy and cooling load makes such changes desirable.

GUIDELINES FOR DESIGNERS

1. Heat loss calculations should be performed using Btu/hr × °F as the basic parameter. The procedure for a residence is as follows:

   a) Calculate the area of each exterior skin assembly having a different R factor, by exposure.

   b) Divide each area by its appropriate R factor, yielding a series of A/R values.

   c) Figure approximately the average difference between indoor and outdoor temperature for the worst 50% of the heating season. Then calculate the difference in temperature between the building and ground, basement or crawl space temperature as appropriate. Multiply the A/R value for components in contact with these more or less constant temperature environments by the ratio of the two temperature differences. For example, if the average indoor temperature is 65°F, the average outdoor temperature is 35°F in mid-winter, and the ground averages 55°F, multiply the slab A/R by $(65 - 55)/(65 - 35) = 0.33$.

   d) If the ground is wet, assume R1. If drained, use R5 for organic soils and R10 for sand or gravel. For basement walls, use outdoor temperature down to a level of one-half the frost-line dimension below grade, and ground temperature below that.

   e) Estimate the air-change rate, multiply it by the volume of air, and multiply the product by 0.018. This yields a number equivalent to A/R. For instance, half an air change per hour in a 14,000 ft^3 building would yield $0.5 \times 14{,}000 \times 0.018 = 126$ (Btu/hr × °F).

   f) Add together all A/R values, plus the equivalent A/R for infiltration. This yields the overall loss rate per degree-hour; multiply by 24 to get Btu's per degree-day.

   g) *Caution:* This is the *gross* building heat loss, before the contributions of window gain and occupant heat are included. If this number is multiplied by the degree-days for your area, and used to predict the required amount of fuel needed, a large error will occur. It is not uncommon in well insulated houses without special passive measures that the *net* load is less than two-thirds the gross load.

For purposes of designing a heating system, the traditional output, showing the worst condition (i.e., the room-by-room loss on the coldest night) can be produced by multiplying the Btu/hr × °F quantity by the

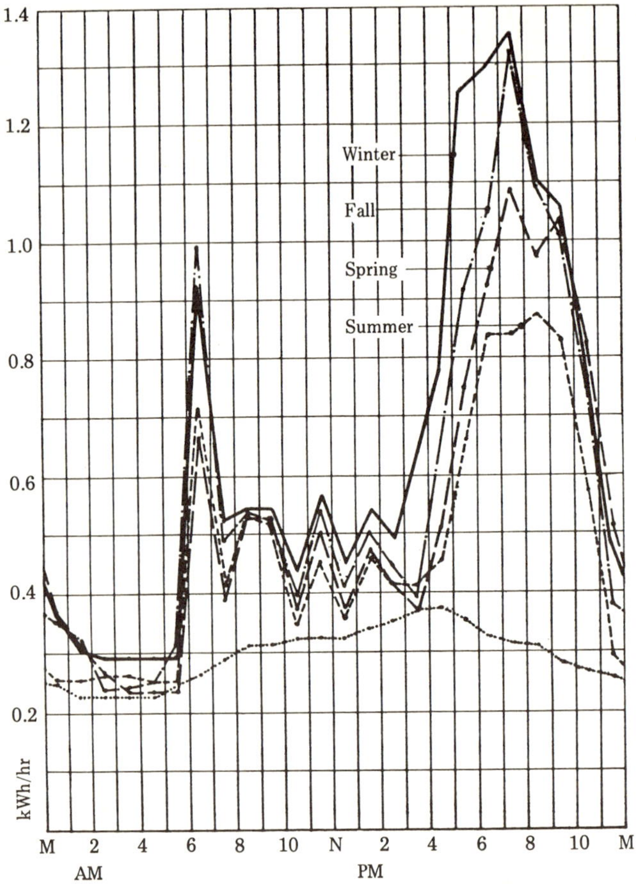

Fig. 6.5 The distribution of the "diversified electric load" by hour and by season for an "average" three bedroom house with a family of four. The diversified load consists of lighting and appliances other than cooking and clothes drying. The plotted values in Figs. 6.5, 6.6 and 8.2 are 90% of the values derived in the General Electric Space Division's study for Sandia Laboratories, "Regional Conceptual Design and Analysis Studies for Residential Photovoltaic Systems," for which Massdesign was a subcontractor. (Document #SAND78-7039, Section 6 of Volume II). The 90% figure is an arbitrary reduction to account for recent dramatic increases in energy costs. Note that the graph plots the energy used per hour, but does not plot instantaneous power levels, which can be much higher than the hourly average value.

ASHRAE design temperature difference for the site. This yields the maximum *rate* of heat loss, or *power* loss.

2. For annual computer simulations, the Btu/hr × °F data are used directly, to find the gross heat loss for each simulation period. From this

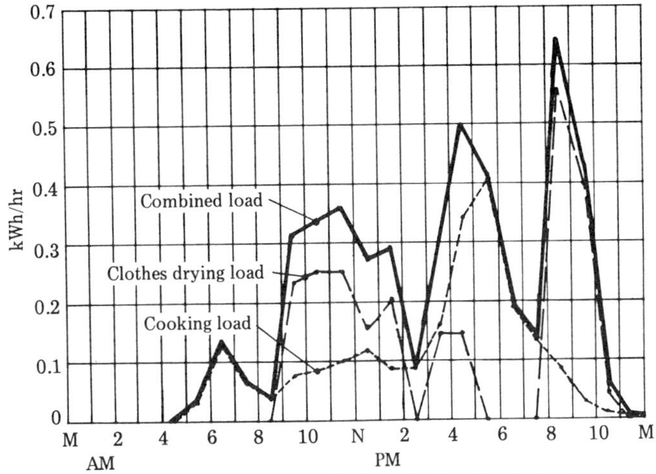

Fig. 6.6 Distribution of cooking and clothes drying loads by hour. These loads are separated from the diversified loads because they can be served either by gas or electricity; or, in the case of cooking, by wood. (Reference Fig. 6.5.)

loss is subtracted all forms of gain: occupancy first, then south glazing, then solar, then heat from storage, finally auxiliary.

3. To account for both loss and gain over time, a calculation procedure is needed which breaks the continuous flow of energy in and out of the system into discrete time periods. A trade-off must be made between accuracy, which leads toward small intervals, and simple computation, which leads toward large ones. Because both solar systems and the buildings they serve are massive and tend to have large time constants, long averaging times are possible. On the other hand, solar gain and occupant-generated heat vary rapidly enough to indicate hourly or even quarter-hourly intervals.

One procedure, used in the author's "Sun-Pulse II" Simulation, is to use two intervals during a day, calculating the behavior during the interval by solving a series of equations describing the various thermal transactions. Such an approach, used with edited weather data, can condense the simulation until it is within the range of a hand-held calculator.

Whether using such a simulation, a larger hourly simulation, or hand calculation, it is necessary to know the value of the occupant heat gain and solar gain through the windows at various times during the day. As noted above, occupant-generated heat varies unpredictably as the occupants come and go, learn and change their behavior, and as children grow and leave home. A useful simulation must allow the occupancy parameters to be *varied*, to see how a given building responds to such changes.

Figs. 6.5 and 6.6 show the results of a recent study, where the average demand of a household for various types of energy was derived from existing studies. An enormous *range* of values found in the existing studies—a sure signal for caution!

Beware of simplified design procedures and obsolete rules of thumb, since they inevitably average in some crucial data which, if varied, will throw the results off substantially. As system design becomes subtler, simulations are likely to become a commonplace tool for the designer, supplementing the convenient but misleading simplified procedures which designers are used to.

4. Be sure to watch for spatial variations in the location of miscellaneous gain. Unless storage or removal is provided, useful energy will be lost either through overheating or through excess ventilation.

7

SUPPLY AND DEMAND: SYSTEM BEHAVIOR AS A FUNCTION OF TIME

TIME VARIATION IN SUPPLY AND DEMAND

In any system where a supply is intended to meet a demand, there is going to be a mismatch. The theory of controls deals with the attempt to make the expenditure of energy meet the task at hand. It is commonly noted that solar energy is an intermittent source—but this begs the question. A furnace is an intermittent source as well. We must dig a bit deeper to understand this problem.

Imagine a constant load, or demand for heat—say a process heating application, where a vat of fluid must be kept at 180°F. One can try to meet this demand by the use of a constant supply, with an output equal to the demand. Experience and probability theory prove the effort impossible. The supply must exceed the demand slightly and be adjustable through a *control feedback loop*, since the demand, however apparently constant, must inevitably fluctuate in time.

One way to keep the vat at 180°F is by turning a burner on and off intermittently. Depending upon the power of the burner, the frequency of firing, and depending upon the *time constant* of the system, we can keep the vat as close as we please to 180°F. The smaller the time constant, the more nearly continuous the supply must be. This is illustrated in Fig. 7.1.

The rate of rise gradually becomes flatter, as the vat gets hotter and loses heat faster. Assuming instant turn-on and turn-off, and a completely homogeneous vat, the temperature oscillations would appear as in Fig. 7.2. Obviously, the scale of the graph is determined by the dead band of the control system (Fig. 7.3). There are other ways to gain control of the supply of energy. We might choose to add together a whole

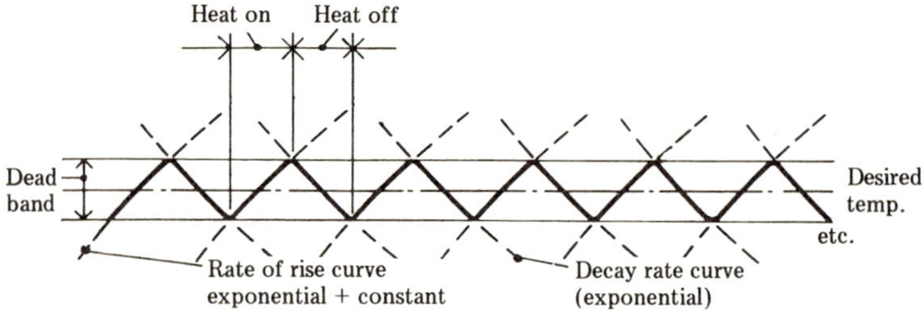

Fig. 7.1 Temperature variation in a constant temperature load being heated by a cycling heating supply.

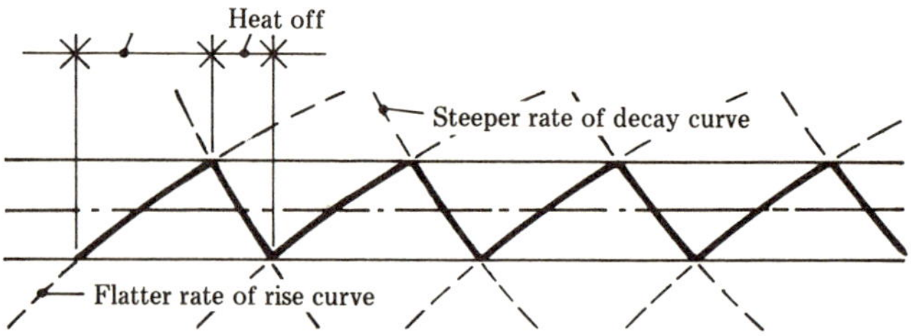

Fig. 7.2 As the load increases, the temperature rises are slower and the drops faster, resulting in longer heating cycles.

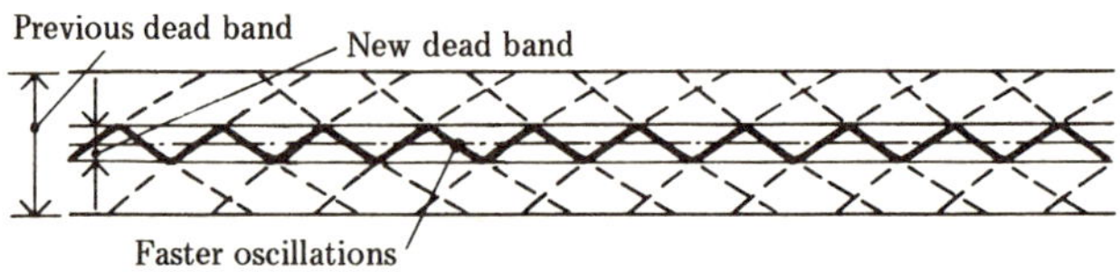

Fig. 7.3 Neglecting the important effects of thermal inertia, the frequency of cycling depends upon the width of the control deadband.

lot of small power sources, cycling them frequently—as in an internal combustion engine; or we might provide a continuous source which can be throttled back at will, as in a gas turbine.

An important variable in deciding how to control the supply of power is the relationship between the cost of the device versus the cost of operating it. In a utility, the issue is crucial. Very expensive generating capacity that is cheap to operate is called "base load" capacity. It is designed to operate on a continuous basis, 70 percent of the time or more. "Intermediate" and "peaking" units operate intermittently at low load factors, are relatively cheap, but cost a lot to operate (typically using fossil fuel). Some of these units are kept idling in what is called "spinning reserve," to cope with very sudden demand surges. With this arsenal of equipment, utilities are able to maintain extraordinary reliability in matching a demand which varies dramatically in time.

A utility builds expensive base load capacity because even at its lowest ebb, demand is substantial. Sometimes, if it is feasible, base load is built to exceed the minimum demand and used to generate excess energy which is stored for later use (for instance, in pumped hydro storage, where water is pumped up hill, stored, and later used to generate electricity when it falls back down hill).

In a building heating system, there is no equivalent to base load capacity in a utility. The plant is relatively cheap, but expensive to operate. This is appropriate, since the heating load is intermittent, reaching a peak on a cold winter night and dropping to zero in the summer. The supply cycles on and off, as in our first example, although there is no reason other than cost why a continuous-fired device with a modulated burning rate could not be used. In larger buildings, several boilers are used, each one being brought on line or taken out of service as the load varies. Electric central heating in residences is often in two or three stages, approximating a modulated-output system.

While space-heating systems typically are sized so that the supply is adequate to meet the worst (= 97½% or 99%) condition, water heaters are usually sized to meet an *averaged* load, with storage intervening to serve peak demands. The storage provides *capacitance* to serve load peaks.

Finally, the load can be "managed," in the sense that, faced with a high energy cost at one time and a low cost at another, the value of using the energy at the expensive time can be reexamined and some energy use postponed until the cheap time.

Summarizing, variations in load can be met by a variety of strategies:

- cycling a fixed output supply
- modulating a variable output supply
- approximating a modulated output with a series of fixed and/or variable output devices
- introducing storage to serve load peaks which exceed the maximum output of the supply
- managing the load to correspond with the availability of supply

The supply devices we have talked about all have *controllable* and *predictable* output. If the device is capital intensive but cheap to operate, it typically is used in a base load capacity, running as much of the time as possible to amortize its high cost. If the device uses expensive fuel, it is typically run as little as possible, to save fuel costs.

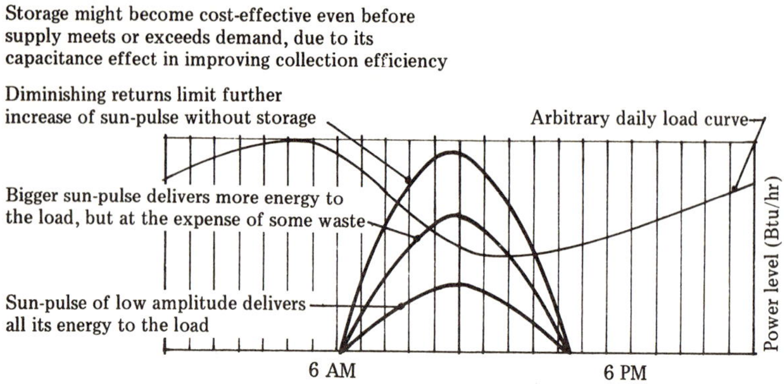

Fig. 7.4 Relationship between the sun-pulse and a typical daily load curve.

With solar, a new element enters the picture: the supply is neither controllable nor predictable.* Faced with this new situation, we notice first that statistically, *some* of the supply is coincident with *some* of the demand. To the extent that this is true, we say the supply is "load following." If the supply power is small enough, a load will "swallow" it, in a sense; that is, the supply power level graphed over time will lie entirely below the demand power level (Fig. 7.4). One also finds that, by *wasting* some of the supply, more of the demand is satisfied. But sooner or later (often even before the supply curve exceeds the demand curve), depending upon the average behavior of supply and demand, and upon the relative cost of storage and collector, it becomes worthwhile to add storage to the system. When we add storage, we are borrowing from one of the control strategies summarized above: we have added capacitance, but this time to soak up excess *supply*, instead of satisfying excess *demand*. It is the difference in function between the two kinds of storage that inevitably leads to separated or stratified storage in solar water heaters—the top satisfies the demand peaks; the cooler bottom soaks up supply peaks. Only when the average supply exceeds the average demand do the two stores merge.

Another control strategy can be borrowed from the first list, namely load management. Whereas load management might normally be used to shift load from a daytime peak to a nighttime valley, it can also be used to displace load to correspond to a peak in the solar output. There is a limit to this technique: we are no longer used to postponing

---

*The peak output is predictable, and with collectors the time of peak output is predictable. With wind devices, only the peak output is predictable. The long-term average energy output can be predicted accurately, but not the power level.

loads indefinitely. The time when solar is available depends on the weather. Even if it were predictable, the weather is uncontrollable. There *are* loads even now which can be saved for available sun; and there will be more, as we learn to live with the new facts of life. The availability of energy "on demand" has become common only in a few countries, and only in this century. Changing our habits may be a very important way of adjusting to a solar future.

MONTHLY SUPPLY AND DEMAND

When viewed as a working system, a solar heated building undergoes a number of superimposed perturbations:

- The daily sun-pulse, jolting the system at a variety of power levels
- The daily swing of ambient temperature and wind
- The daily changes in thermostat setting
- The daily cycle of occupant-generated heat
- Variable, several-day-long swings in ambient temperature and wind as weather fronts move across the site
- The regular, annual cycle of the clear-day sun-pulse
- The regular, annual cycle of ambient temperatures

Because of the variable time periods of these changes, it is impossible to characterize the system by a daily, monthly, or annual cycle *alone*. The system responds differently to the supply and demand, depending upon its current state, which in turn depends upon the cumulative effects of all these changes acting together in previous states.

The system does not simply use solar heat to satisfy the load and salt away the excess for later use—if it were that simple, nomograms would have been developed long ago for system designers to use. Rather, a balance must be achieved by the designer among the three key components in the system—collector, store, and load heat exchanger—which responds to the continually changing conditions occurring through the swing of the seasons in a way which optimizes the *annual yield* of solar energy.

By far the most important element in the system is the collector. Its size relative to the load overwhelms all other system variables and deserves study separately. If a collector is tilted at an angle about equal to its latitude, its potential output will drop off noticeably in midwinter and a slight amount in midsummer. While the output of a collector is a function of the system as a whole (and in particular, the store temperature), as a first approximation it is fair to graph the potential output as if the collector were run all year at a fixed ΔT—say, 55°F. We can then

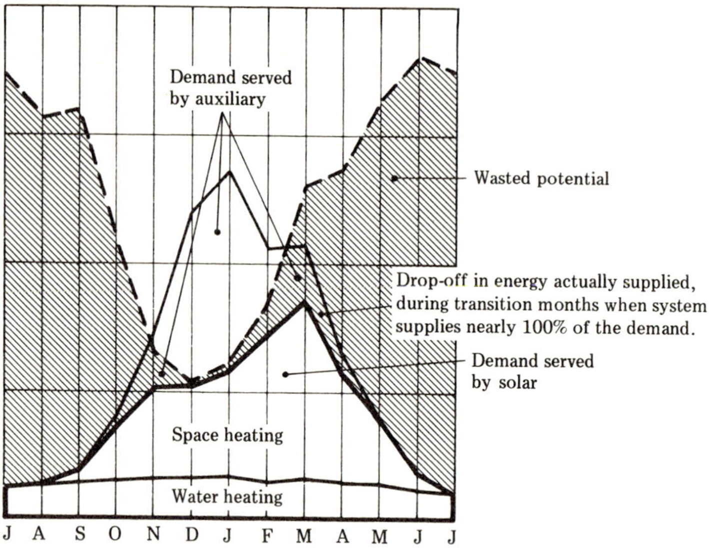

Fig. 7.5 Superimposing the potential supply of a solar heating system (figured with a constant temperature difference—a reasonably good approximation for many systems) on the monthly load curve. In the transition months, an averaging effect occurs which prevents 100% solar until there is a substantial excess of supply above demand. The "wasted potential" is dissipated through collector stagnation, high collector losses and high store skin losses.

compare the potential output with the demand, as in Fig. 7.5. During midwinter, the demand is so large that the output is almost totally used; but in the "transition" months, when the potential output of the system approaches and then exceeds 100 percent of the demand, there are occasions (usually) when the system overproduces, followed by occasions when it cannot keep up with the demand. Even though the average output equals the average demand over the month, on a shorter averaging period the two are unequal. In the graph, the dotted line represents schematically the falloff in usable output during the transition months. Otherwise, the actual output equals the lesser of the potential supply or the demand and is shown in the diagram by the shaded area.

Manipulation of the other system variables, such as store size, controls, and load heat exchanger, as well as the selection of various collectors, can change the exact quantity of useful output. The schematic outline remains intact, and many useful concepts can be drawn from such a graph.

In these graphs, the space heating load referred to is the *net* load. The gross load is the loss from the house or building; the net load is the gross load minus the supply from miscellaneous sources of heat and

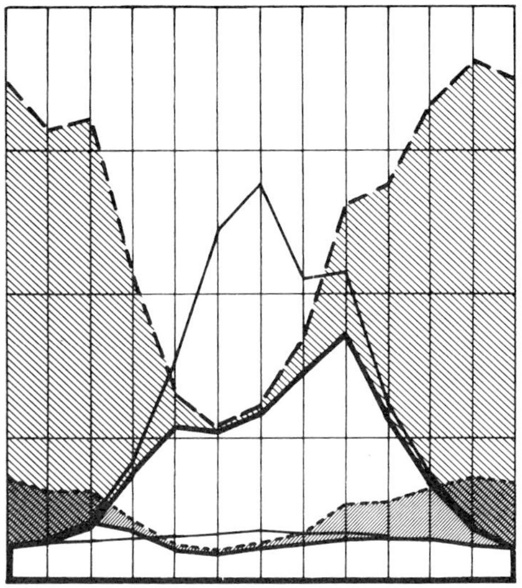

Fig. 7.6 Because the supply and demand curves are similar, water heating systems have less wasted potential (lower curve).

from incidental solar gain. In a deliberately passive solar house or in one using extraordinary energy conservation measures, the net load is quite small; here we refer to houses with a moderate amount of solar gain-perhaps no more than 15 percent of the house area in south glass.

First, notice what happens when the size of the system increases or decreases (Fig. 7.6). For a given load, a larger system tends to *waste more potential*, while a smaller one wastes less. On the scale of a house, smaller systems are more expensive on a unit-cost basis than larger ones because they must amortize fixed costs over a small collector area. Within a remarkably wide range the two effects balance to produce a relatively constant cost effectiveness. For large loads a "small" system —that is, one satisfying a small fraction of the load—may be quite large, so that economies of scale are already present, making the smaller system noticeably more cost effective than a larger one which has substantial wasted potential.

Next, notice the difference between a hot-water load and a ventilation or space-heating load (Fig. 7.7). The former is more or less constant from month to month, varying with the temperature of incoming water; the latter peaks in midwinter and, except in the coldest climates, falls to zero in midsummer. The ratio of the two loads is very important for a solar heating system. As the hot-water load increases relative to

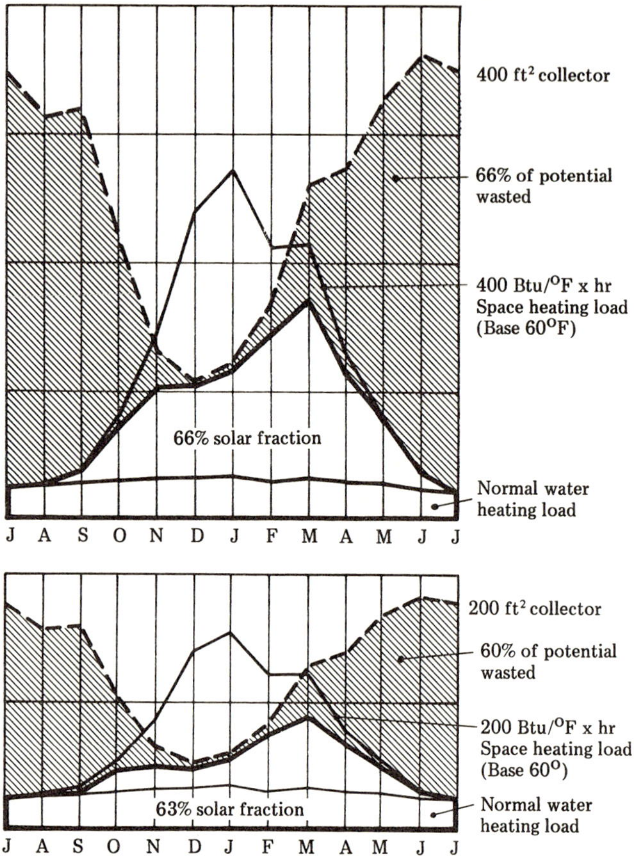

Fig. 7.7 As the water heating demand becomes larger relative to the space heating demand, the demand curve flattens, and a system delivering the same solar fraction suffers less wasted potential. In the example, the space heating load was cut in half, as was the size of the system; but the same principle would apply if the hot water load had increased proportionally.

the space-heating load, it is easier to amortize the solar equipment, and, all other things being equal, a given system will show a greater cost effectiveness.

It is easy to see that you can draw diagrams like this for a large variety of combinations. The interesting point is that many decisions can be made on the basis of the diagrams:

1. It is always better to pick up the hot-water load along with the space-heating system, because this strategy extends the system's period of operation and increases its cost effectiveness.

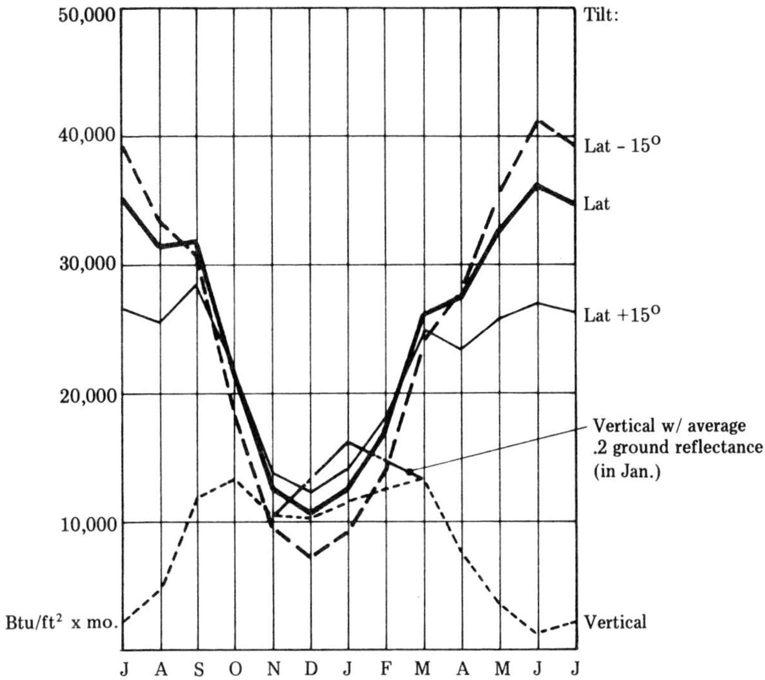

Fig. 7.8 Potential monthly output curves for a collector at various tilt angles. Unlike many similar graphs, this one takes account of off-angle reflections and operating losses (a single-glazed collector operating at a constant ΔT of 55°F was simulated, using Sun-Pulse II/TMY data from the author's simulation).

2. With a given load, system cost effectiveness will increase or decrease depending upon whether, as the system gets larger, its unit cost goes up more slowly or faster than its output.

If you look at the potential output curves of a collector with a variety of tilt angles (Fig. 7.8), you will immediately notice that a steeper tilt optimizes for space-heating loads, and a shallower one for hot-water heating. So, within certain limits, it is possible to "shape" the output curve to follow the load. These limits are severe:

1. Shading becomes more of a problem for steeply tilted collectors, because they get more of their energy at low sun angles.
2. Shallow tilts can cause trouble in snowy climates, because snow may tend to stick to the collectors—35° seems to be a safe minimum for most such climates.
3. The output during the transition months often constitutes a large percentage of the total annual harvest.

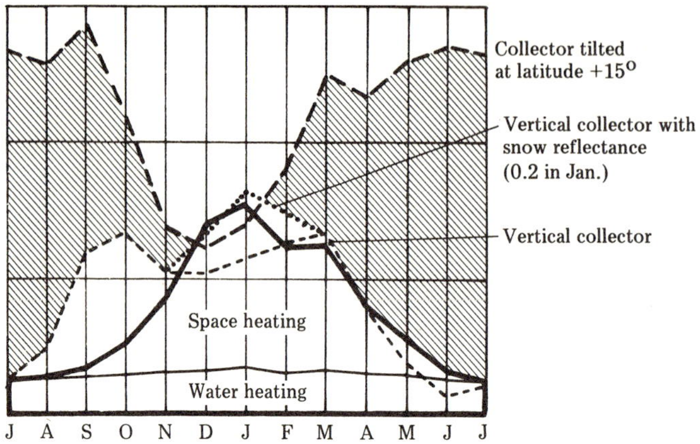

Fig. 7.9 Supply and demand curves for a nearly 100% system. Because the vertical collector follows the demand curve so closely, there will be a continual shortfall due to averaging effects. The tilted collector will deliver 100% of the energy for most of the year. The apparent "waste" of the tilted collector reflects its much higher yearly potential. The lack of frequent stagnation can be an advantage for the vertical collector.

Steep collectors are not as good during these times. By emphasizing output in the transition months, one can keep the entire auxiliary system off for several additional months in spring and fall—a helpful strategy for fossil-fuel auxiliary systems.

Finally, these diagrams explain why 100 percent solar space heating with short-term storage is such an elusive goal. Instead of an average load, one must combine the worst output curve with the worst load curve. Then the load curve must exceed the output curve by some amount, in order to insure that an adequate averaging period is achieved (the same phenomenon that occurs during the transition months for a normal system). The diagram is shown in Fig. 7.9. The waste is large, because the system is working to capacity only once every few years, while the rest of the time it is oversized.

Supply/demand diagrams can be put to yet another good use. It is not widely known that the effects of energy conservation by insulation and by reducing load temperature have a different effect upon the load. (See Fig. 7.10.) Insulating does not change the width of the gross load curve, but only its amplitude. Reduction in temperature by reducing thermostat settings not only reduces the amplitude of the peak, but shortens it as well. (The difference between the two is diminished somewhat by the fact that, in the first instance, occupancy heat and solar gain satisfy the load farther into the spring and fall "shoulders" of

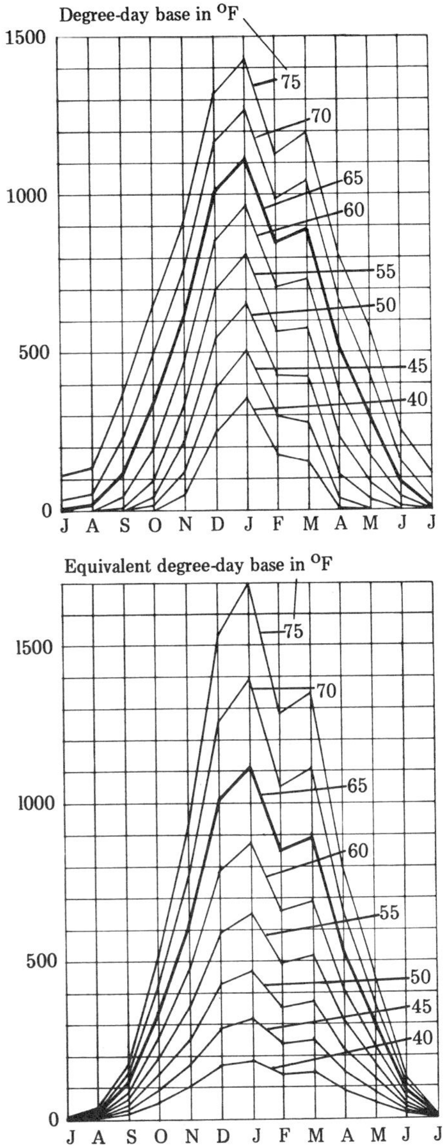

Fig. 7.10 A comparison of two energy conservation strategies. The left-hand graph plots the Boston TMY degree-day load to various bases. The right-hand graph plots the monthly loads which result from changing the building heat loss rate for the base 65 degree-day condition, to yield the same annual heat loss as in the left-hand graph.

the curve.) All other things being equal, the load temperature reduction strategy leaves the solar system with more of its potential wasted than does the insulation strategy. Perhaps more important, comparing the percentage of the net load satisfied in midwinter by two equally cost-effective systems, the one serving the cooler house will be lower than the one serving the warmer but better insulated one. In general, a short heating season is less satisfactory than a long one as a solar load, illustrating one example of the competition between solar heating and various energy conservation measures.

SHORT-TERM SUPPLY AND DEMAND DURING NEARLY 100-PERCENT SOLAR PERIODS

In the preceding diagrams, the output and demand for an entire month were shown. It was noted that when the system came close to providing 100 percent of the demand, some of its output was wasted, and that this condition occurred for all systems during the transition months, i.e., those in which the supply of solar energy reaches and then exceeds the demand.

If a closer look is taken at smaller averaging periods than one month, new insights can be gained, but we must begin to take account of system variables other than the collector size. Fig. 7.11 compares two identical systems, one having twice as large a store as the other, during part of a transition month.

Let us suppose that over a six-day period the potential output of the system (based on the assumption of a constant 55°F $\Delta T = T_i - T_a$)

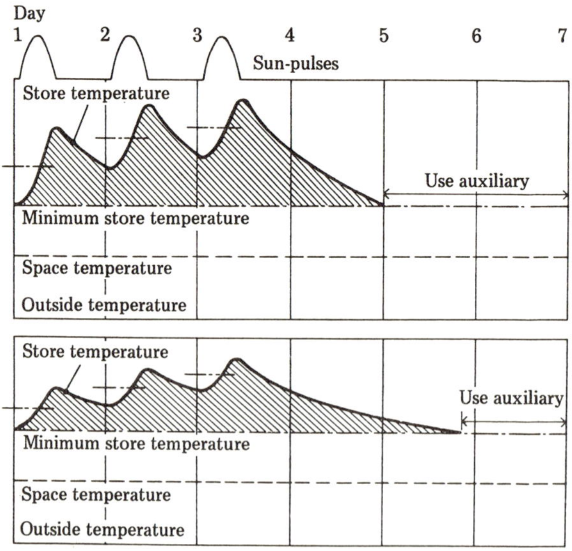

Fig. 7.11 Comparison of storage temperature in two systems differing only in the size of the store and load heat exchanger, for three sunny and three cloudy days in succession. More energy is collected with the larger store. Store skin losses must be held equal.

exactly equals the net load to be satisfied. Let us also suppose that the first three days are sunny and the next three cloudy, and that outside temperature stays close to the monthly average throughout the period.

The systems will behave as follows. On the first day, both will accumulate a surplus of energy, raising the store temperature. On the second day, the system with the larger store will collect energy at a lower temperature and thereby collect more energy; and so with the third day. The total energy accumulated by the larger-store system will be greater, the amount of difference depending upon how sophisticated the collector is (both systems, for comparative purposes, are assumed to have the same collector array). If the array is, say, a vacuum tube design, the difference between the two systems will be very small; if it is a single-glazed nonselective surface design, it will be quite noticeable. This effect results from the increased losses from the collector, losses which are proportional to the collector's "U" factor—small in the vacuum tube, large in the low-tech collector.

Now, it is clear that during the three subsequent days the system with the larger store will provide more of the demand than the one with the smaller store. Both have plenty of time to discharge, both are equally affected by the low-limit control. But the system with the larger store, simply, has more energy to deliver, and so will do better. Accumulated over the entire transition period, such differences can make a noticeable difference in annual system output. As the store gets larger, the load heat exchanger must be resized as well, as discussed in Chapters 5 and 9.

In the two previous examples, three sunny days were followed by three cloudy days. What is the effect of different patterns of weather during the 100-percent solar transition months? It is not hard to determine that a regular cycle of "sun-on/sun-off" yields a better energy harvest than long periods of sun followed by long periods of clouds (Fig. 7.12). This effect is very important for systems trying to supply 100 percent of the net load under the worst conditions in midwinter. The system, if tested against average weather data, will optimize with substantially less storage and load heat exchanger capacity than if tested against long spells of cloudy weather. Although the term "carry-over" has no meaning in a normal, transient system (under what conditions would it be defined?), it could be applied to a nearly 100-percent system and related to the system's storage and heat exchange capacity. When storage becomes large enough to carry the full space heating load under the worst conditions for several days, the normal relationships between collector and store vanish. Whereas a "normal" system might have between 2 and 5 gallons of storage per square foot of collector, Mark Hyman's nearly 100-percent house near Boston had a ratio of 13

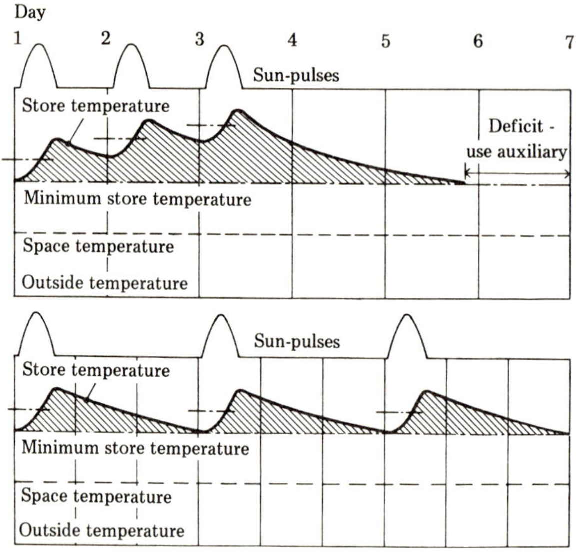

Fig. 7.12 The regular alternation of sun/no sun days yields a better energy harvest than long sequences of identical days. This factor must be accounted for in simulating system performance during periods of nearly 100% solar fraction.

gal/ft^2. The load heat exchanger must also be larger, enough to supply the load at a reasonable storage *discharge* temperature. (If the system is to be 100 percent, it must be so at the very end of the cloudy period, and under the worst ambient temperature conditions.)

PHASE-LAG (SEASONAL) STORAGE

In a normal solar heating system, the store functions partly as a capacitor, to dampen oscillations in the system. In particular, storage is used to reduce the operating potential at the collector, reducing losses and thereby improving efficiency.

There is another, more familiar role for storage, that of "banking" energy for later use. Naturally, in its role as a capacitor, storage must bank energy over the short run; but an ordinary store cannot bank energy over a time period larger than the period of its operating cycle for the particular time of year involved. To "carry over" energy for longer periods requires much larger stores.

Consider the typical supply and demand curve shown in Fig. 7.13. As noted earlier, a good deal of the potential supply is wasted, and there is a short-fall in midwinter. Fig. 7.14 shows the same graph plotted as a surplus or deficit compared to the demand. Notice that the potential

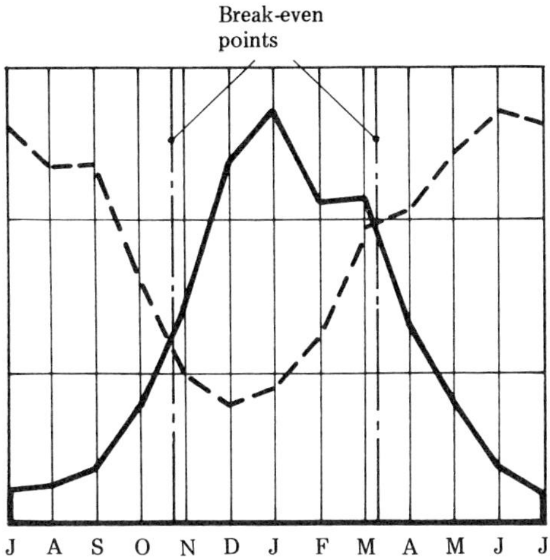

Fig. 7.13 Typical supply and demand graph for an energy-conserving residence with a tilted collector.

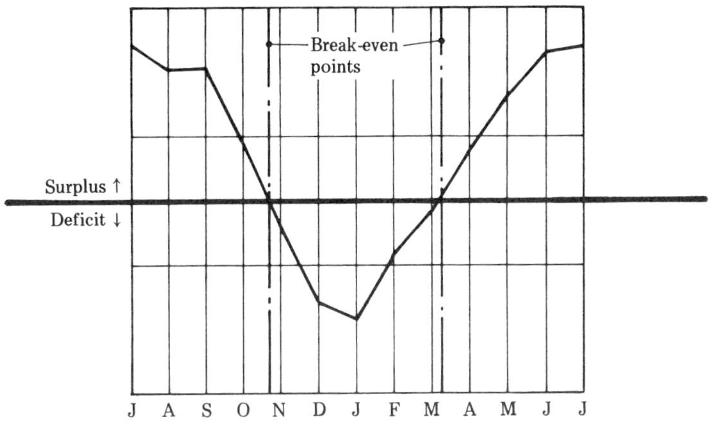

Fig. 7.14 Fig. 7.13 replotted as a surplus/deficit curve relative to a horizontal break-even line.

supply curve is a wave form, peaking in midsummer and reaching a nadir in midwinter.

With a normal (25 Btu/°F × ft^2 collector) store, the primary effect of a surplus of energy is an increase in losses, both from the store itself and from the collector. These losses set a natural limit to the store temperature for any given energy surplus. Most systems include a high limit cutoff, since the natural high limit is often above the rated temperature of some key system component. A secondary effect of a surplus is an increase in stored energy.

Suppose that the store is very large relating to the supply, and that

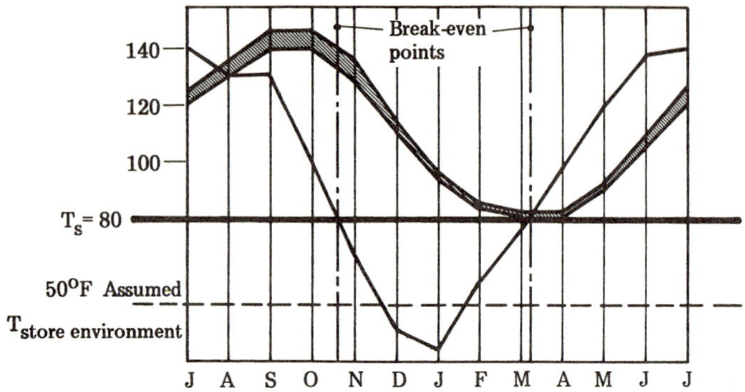

Fig. 7.15 If the store is large enough and the skin losses small enough, the store will accumulate energy during the surplus periods for discharge during deficit periods.

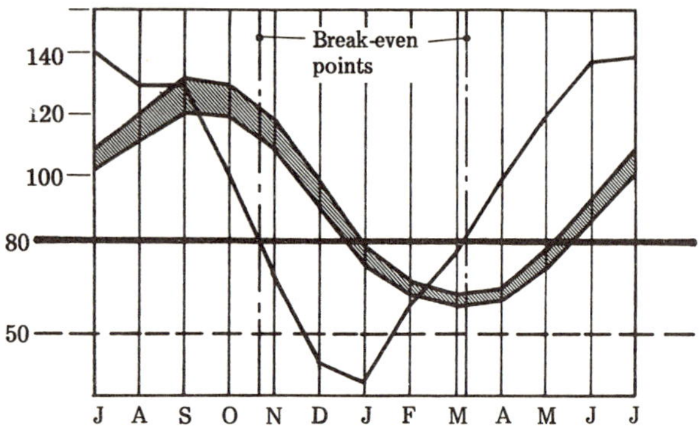

Fig. 7.16 Modest increases in skin losses, accumulated over an entire season, can ruin the performance of a seasonal phase-lag system. Here the skin losses have increased 50%.

storage skin losses remain small. The primary effect of an energy surplus would now be an increase in stored energy. The collector would continue to operate at a low temperature due to the very slow rise in store temperature. Fig. 7.15 shows the seasonal rise and fall of a very large store with very small skin losses, for the system graphed in Figs. 7.13 and 7.14. If skin losses are zero, the peak and minimum store temperatures occur at the seasonal "break-even" points shown in Fig. 7.14. Store losses cause the peak to occur earlier in the fall, but have little effect on the spring nadir.

With the postulated small skin losses and very large store, the solar fraction provided by the system has risen from 65% to over 95%. It is not 100%, since the low mid-winter store temperature, adequate for

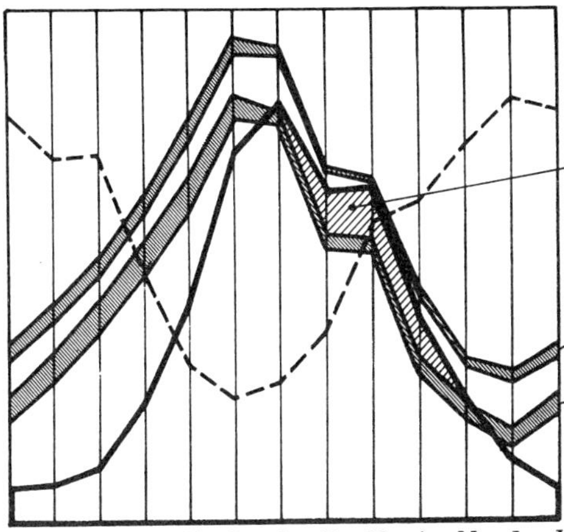

Fig. 7.17 Figs. 7.15 and 7.16 replotted in the normal way.

space heating with a suitably large load heat exchanger, cannot fully satisfy the water heating demand.

The very large store with a very small skin loss rate provides sufficient capacitance to cause a "phase-lag" in the seasonal energy curve, accumulating energy during the warm months and discharging it in mid-winter. In Fig. 7.16, the skin losses have been multiplied by 1.5. While the phase-lag remains, the solar fraction drops noticeably. The crucial influence of skin losses is clearly shown in Fig. 7.17, in which Figs. 7.15 and 7.16 are replotted in the normal manner.

In summary, as long as the integrated area under the supply curve (reduced by the appropriate store losses) exceeds the area under the demand curve, with the store maintained at a high enough temperature to satisfy the load, it is theoretically possible to deliver 100-percent solar heating with a modestly sized collector. (Water heating can be handled by the use of a heat pump, to hold down store temperatures while delivering 70 percent or so of the demand.) But the control of skin losses is absolutely critical. It is remarkably easy to design so-called annual-cycle or seasonal-storage systems which do not achieve the goal of long-term carry-over, but are merely elaborate mechanisms for heating ground water or outside air.

SYSTEM CONTROLS I: LOW LIMIT ON DELIVERY TEMPERATURE

Besides the *sizing* of components, the *control system* can have a profound effect upon the system operating temperature, and thereby upon the annual system output.

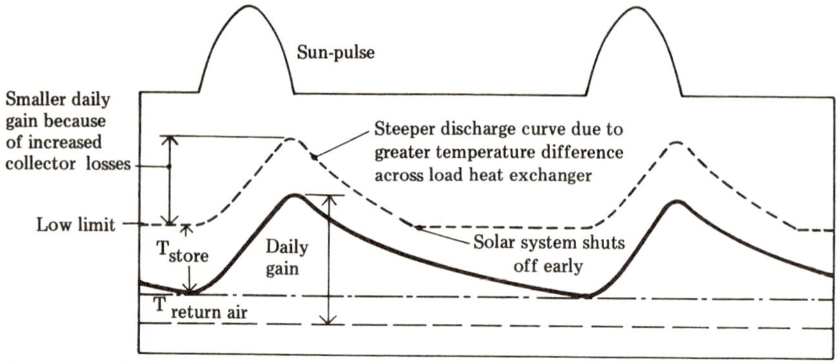

Fig. 7.18 Imposition of a low-limit on storage discharge always degrades system performance.

It is very common to install a low limit on the output of the load heat exchanger. While the problem will be discussed in detail in Chapter 10, the effect of such a strategy can be dealt with in general terms here. (The discussion does not apply to 100-percent solar systems.)

Our goal is to operate the system at the lowest economical temperature. By enlarging the store, we hold down collector temperature. To compensate for the resulting increase in system discharge time constant, we enlarge the load heat exchanger: the large heat exchanger drives down the store temperature overnight; the large store keeps the temperature down during collection.

Suppose a low limit temperature to the output from storage is added—typically, it would be set high enough to avoid discharging air to the space at a temperature which feels cool to the occupants—values of 75°F to 85°F are commonly used. It should be added that higher low limits—100°F to 135°F—have been used by system designers in the mistaken belief that lower temperatures were inadequate for system operation.

By setting a low limit on the discharge temperature, the system is held at a higher temperature throughout its cycle. No matter how large the load heat exchanger is, the system cannot discharge below the low limit. Obviously, the first step is to design the distribution system to eliminate or minimize the low limit. But if this is impossible, the question arises: what is the best way to deal with a low limit?

Fig. 7.18 shows a system with and without a low limit. The system is optimized without a low limit, with both the amplitude of the rise in store temperature, and the temperature difference between return air and storage at the beginning and end of the cycle reduced to their economic minima. (We have in effect bought the lowest temperatures we could afford.)

Notice these effects caused by adding the low limit:

1. The amplitude of the store temperature rises, and hence the amount of energy added to the system is reduced due to the higher operating temperature of the collector.
2. The load heat exchanger discharges faster during collection, because of the higher temperature difference between store and house, to some extent *reducing* the average operating temperature and compensating for the rise (but only on the coldest days—on most days, there is little load during collection).
3. The second effect never compensates for the first, and system output will go down.

A larger and more expensive store will become optimum, raising the cost of the system. Cost effectiveness will then have gone *down*, since more money has to be spent on the store to maintain the annual output. Leaving the store the same size will likewise reduce overall cost effectiveness. Setting a low limit, while sometimes necessary, always reduces cost effectiveness.

The problem can be avoided in certain kinds of systems by a control circuit which turns on the solar *below* the low limit whenever the backup system is on. Whether such a control is cost effective is a matter for individual calculation. The author sometimes uses it, especially if such a circuit can be included in a single packaged controller, but prefers a simple owner-operated switch to defeat the low limit whenever no one is at home, or at bedtime. All of these ideas depend upon a backup which heats the house air *after* the solar system has first applied some preheat. This would typically exclude hydronic systems operating at temperatures above that of the solar storage tank, discussed later in this chapter.

With baseboards or other heat exchangers relying on passive convection to drive the air through the exchanger, a "natural" low limit occurs, but one which "floats" about 20°F above the air temperature at the inlet to the device (as opposed to a fixed low limit which does not change with thermostat setting). Fig. 7.19 shows the output curve of a typical baseboard convector.

SYSTEM CONTROLS II: NIGHTTIME THERMOSTAT SETBACK

Following the principle that every energy conservation measure should be taken regardless of whether solar heat is used, the homeowner or

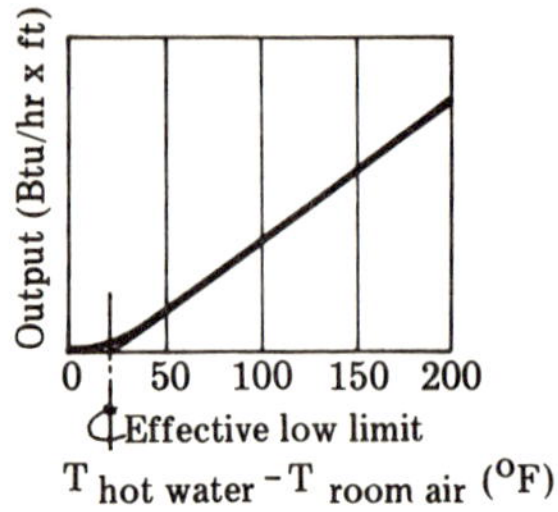

Fig. 7.19 The output curve for hot-water baseboard units is roughly linear from a temperature about 20°F above room temperature. Baseboards are an excellent output device, if long enough, because they are quite cheap to install.

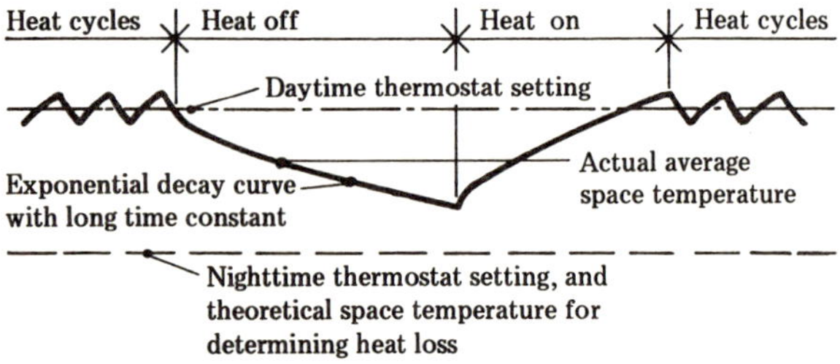

Fig. 7.20 Space temperature graph of a hypothetical massive building. While the decay curve could represent either a massive or a very well-insulated building, the long warm-up period only occurs in massive buildings. Notice that the nighttime thermostat setting was never reached.

building operator should always set the thermostat back at every opportunity. What happens to a simple solar heating system when the thermostat is set back?

Earlier, we emphasized the effect of building mass on its thermal decay. Before considering nighttime setback, decide whether it is worthwhile. If the space has a very large thermal capacitance, it simply won't cool off very fast, as shown in Fig. 7.20. If the energy supply is turned off, it may coast for several hours. To get the building back up to temperature for the next morning, the backup system might have to be turned on several hours before occupancy. So, to experience the advantages of nighttime setback, as much of the building's storage capacity as possible should be separate from the space, in an insulated container. The less intrinsic capacitance, the faster the space will cool, and the longer it will remain at the lower temperature, saving money for the owner.

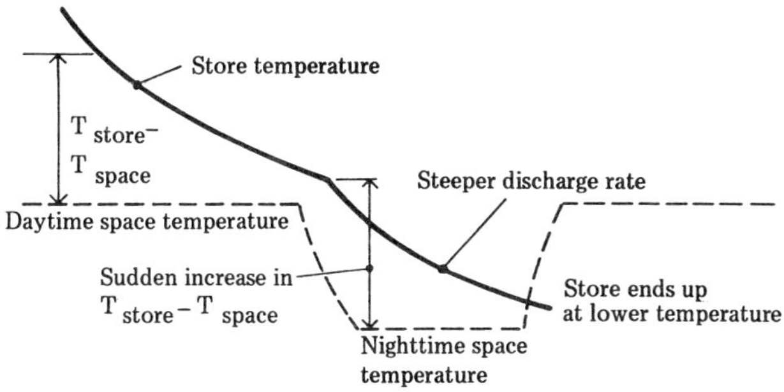

Fig. 7.21 Graph of store temperature in a solar heating system discharging continuously to a space, at nighttime thermostat set-back. Collector operates well the next day.

Given an active system in a low-mass building, Fig. 7.21 shows what happens to the load heat exchanger when the thermostat is set back and the building cools. (Assume for a moment that there is no low limit on discharge temperature.)

The discharge rate from storage is based upon the temperature difference between the water in the store and the return air temperature. When the building cools, so does the return air temperature. The load heat exchanger "sees" a larger temperature difference and begins to work harder, along a new and steeper decay curve. By the time the night is over, the system will have dropped to a lower temperature than in the first example in which the thermostat setting was constant, with a corresponding improvement in the next day's collector operation. In fact, it might be at a temperature lower than that of the daytime thermostat setting. If there is no low limit on the system, and there is no collection the next day, energy will feed back into the store from the return air through the coil. This process is, surprisingly, not particularly wasteful (it is not wasteful at all if all pump and fan energy is lost to the heated space), and it is thermodynamically interesting. The building, in effect, is being used exactly like a solar-assisted heat pump, to drive down the store temperature.

Low limits will inhibit the beneficial effects of nighttime setback unless the low limit is overridden during backup operation, or by an automatic or manual switch when the occupants are not inconvenienced (as at night, or when the building is vacant). In hydronic systems, no low limits are necessary, and nighttime setback can help overcome the 20° minimum temperature needed to drive the hydronic system. For example, in the author's design for Drumlin Farm Nature Center, the store habitually reaches 75°F by morning—the nighttime thermostat setting of 55°F plus 20°. On a sunny day, the 15° setback (from 70°F to 55°F) will save (collector loss rate = 1.25):

$$15°F \times 7 \text{ hrs.} \times 1.25 \text{ Btu/ft}^2 \times \text{hr} \times °F \times 840 \text{ ft}^2 = 110{,}250 \text{ Btu}$$

A simple two-stage control system exhibits an unfortunate quirk during the spring and fall. At these times, the solar system is often capable of serving the entire building demand. During the day and night, only the solar load heat exchanger cycles, and the second thermostat contact is never actuated. But when the thermostat is turned up abruptly in the morning, the backup system will come on willy-nilly, no matter how hot the solar tank is. It is sometimes cost effective to install a high-limit relay, sensing storage discharge temperature, which prevents the backup from coming on above an adjustable limit. Alternatively, a simple time delay can keep the backup off long enough to let the solar system satisfy the load.

SYSTEM CONTROL III: CONTROLLING THE BACKUP SYSTEM

The number of ways in which the back-up energy supplies can be woven into a solar heating system are truly mind-boggling. Since every possibility cannot be discussed, two examples have been chosen to show how the thermodynamic principles outlined in this book can be applied in specific instances.

The simplest way to distribute heat from a water store is to place a coil in the airstream of a warm-air heating system. The backup can then be an ordinary furnace, the indoor section of a heat pump, electric resistance heat, a coil fed from a boiler, or any combination. Like a utility, we first use energy from the "base-load plant"—the one which has the highest capital cost and the lowest operating cost—namely, the solar-generated energy. Then we use energy from a plant intermediate in both capital and operating cost, from a heat pump, for example; and we top off the supply with low-capital-cost, high-energy-cost electric resistance heat. So the first rule is to place the solar coil in the return airstream in front of the furnace or of any other coils.

The best control sequence uses a two-stage thermostat. Both stages are wired to turn on the solar coil pump and the fan; only the second stage turns on the back-up energy source. If a low limit is present, it is wired to defeat the first stage. Similarly, a high limit or time delay, to prevent back-up operation at thermostat set-up, can defeat the second stage.

A second example illustrates more sophisticated issues which arise in the control of backup systems. Suppose we wished to tie a solar supply into an oil- or gas-fired hydronic system. The first step is to make the hydronic system as cost effective as possible. One way to do this would be to link the boiler temperature with an outdoor reset, to minimize standby losses. The theory here is that there is no reason to operate a boiler at a temperature higher than necessary to satisfy the load, plus some standby capacity to help bring the house temperature up in the morning. (As noted earlier, baseboard units and other hydronic devices have outputs which are roughly proportional to the difference between their operating temperature and a temperature at or somewhat above room temperature.) Let us say we have 20 percent reserve capacity. If the boiler temperature is reset accurately, the circulating pump would on the average operate 80 percent of the time. The boiler would of course cycle more or less frequently, depending upon its temperature; but on the average, the stack losses would be less with lower temperatures, and the burner nozzle could be kept as small as possible. A flue damper, if a safe and approved one can be installed, will also help reduce system losses; but the two measures are supportive, since both are very cost effective. (If neither were very cost effective, you would not want to do both, since one would undermine the cost effectiveness of the other; see the argument later concerning solar and heat pumps.)

Now, what is the best way to link solar heat and hydronic heat? By analogy with the first example, we would like first to heat with solar, and then to heat with the backup. In Massdesign's Drumlin Farm Nature Center retrofit for the Massachusetts Audubon Society, two separate hydronic systems were installed—one for solar and one for backup, making the "both/and" control system work. But in the present example, we wish to use the *same* hydronic system. Either we must use the solar only when it is warmer than the backup; or we must use it in between backup cycles, in an "either/or" mode of control. If we had not first optimized the backup system, we might well have used the either/or control. Only on a few occasions does a hydronic system with a constant temperature setting work to capacity. The rest of the time, the circulator cycles, being off more than half the heating season. There would be plenty of "in-between" times for the solar-heated water to circulate, even allowing for a cooling-off period after each hydronic cycle.

But, wisely, we first optimized the backup system. (We should *always* optimize the system before applying solar. This is not good for business, since often the solar proves uneconomic if it can't claim credit for optimizing the original system.) By so doing, we make the backup circulator run nearly all the time, leaving too little time for the solar to discharge its accumulated heat. This raises the collection temperature and forces us to use a more expensive collector, and to insulate the store much more thoroughly to compensate for higher store temperatures.

The next step would be to increase the reserve capacity of the hydronic system by raising its temperature, so that the circulator would operate 70 percent, then 60 percent of the time. Each time, we would examine the implications for the solar system, to see if the *overall* system cost effectiveness improved. The goal here would be to sacrifice some efficiency in the backup system to allow the use of more cost-effective collectors working with storage cooled during the now-lengthened periods between backup cycles.

Using another approach, it might prove cost effective to increase dramatically the length of baseboard, and thereby reduce the required system operating temperature of the backup even more. This would allow the simultaneous use of a moderately cheap collector and the backup system, working exactly like a warm-air system off a two-contact thermostat, in the "both/and" mode.

In both these examples, solar and backup are both controlled by a single thermostat. If electric baseboards with integral thermostats are used as backup, it is possible for the solar thermostat setting to be lower than that of the backup thermostat. This would prevent the operation of the solar load heat exchanger, and energy would accumulate in storage; the system would then be zero percent solar. The problem can be prevented by controlling the baseboards centrally, thereby losing some of the flexibility of the baseboards (individual room control); but the alternative is worse.

STRATIFICATION IN WATER STORAGE

In an operating collector, glazing losses and absorptivity determine how much of the impinging energy "gets through" into the system; its temperature relative to ambient determines how much heat is lost to the environment; and a combination of its effectiveness as a heat exchanger and the heat capacity rate of the fluid passing through it determines how high a temperature rise is necessary in the fluid to remove the net available heat. Since we want to express the output of a collector in terms of the incoming fluid temperature, the term F_R is used to measure the combined effectiveness and heat capacity rate of the collector.

An "ideal" collector experiences no rise in fluid temperature. One can do no better in holding down losses than to operate at the load temperature. Any temperature above that will increase losses. It follows that within reasonable cost limits the designer should strive to increase the heat capacity rate, thereby increasing the F_R and holding down losses.

Looking at the store, it is usual to withdraw the fluid from the bottom or one end of the store (be it rocks or water) and return it to the top or opposite end. Why is this? Some say it is to *encourage* stratification in the store. It might better be said to *take advantage of the inadvertent stratification*. Clearly, if it is desirable to reduce the temperature difference in the collection fluid between the inlet and outlet to the collector, it cannot be desirable to encourage stratification in the store, since that requires the opposite strategy, namely *increasing* the temperature difference. It is sometimes hard to realize that in our simple system there is no other way to stratify storage than by using the available difference coming from the collectors. The *load* is not going to increase stratification, it is going to decrease it, by withdrawing energy and thereby reducing the available potential for stratification. And measures which involve heating the store or cooling it sooner or later reduce collector efficiency or require almost as much outside energy as they save. Unfortunately, one very effective method of increasing stratification in storage tanks is quite commonly used; namely, undersized or wet insulation at the bottom of the tank. The author hopes that the futility of such a measure goes without saying.

In a drain-back system, the return water stream can stir the tank and partially offset the stratification which would occur. There are various ways to prevent this, all involving some means of slowing down the water entering the tank. The discharge pipe can be increased in size several feet from the outlet; a perforated screen can be added below the outlet to prevent stirring; or a soaker can be added at the end of the discharge pipe to break the discharge into small, slow-moving streams. The cost effectiveness of such devices is questionable with systems having very small temperature rises, but should be investigated. Avoid anything which increases maintenance.

Suppose as is commonly the case that a minimum temperature is set below which the storage tank is not allowed to discharge to the load. One might think that if the tank were artificially stratified, more energy could be delivered to the load than otherwise, since more of the tank would be above the minimum delivery temperature. Some reflection will show that this could not be true. Consider such a system in an equilibrium condition—constant outside temperature and daily sunshine of the same magnitude each day. The system will settle down into a pattern wherein the store discharges all the energy picked up during the day by the next morning. This must be so, or else the store temperature would rise and the system would not be in equilibrium. In this stable state, the discharged condition would find the bottom of storage at some temperature between the minimum delivery temperature setting and the temperature of the return airstream (or the return water stream, if a hydronic system) going back to the space to be heated. The top of the storage would be at the minimum delivery temperature. Nothing could be done to the system to produce lower temperatures in the storage tank. If the flow rate through the collectors is slowed down in order to increase the initial stratification, less energy will be collected, and less will be delivered to the load.

"Improvements" can be made to the simple design discussed in this book, which will enhance stratification. These involve either increasing the size of storage so that it "turns over" only once a day; or introducing a coil on the inlet side of the collector pump to avoid turning the store over at all. These are refinements which are seldom used in real situations, but might be considered by the designer.

ROCK STORES

Throughout the book, liquid-cooled collectors and water stores have been used as the example of simple, cost-effective solar systems. Obviously, the proponents of air systems and passive systems will find this bias, if it be one, puzzling if not irritating. Repeating the reasons, the basic thermodynamics of the liquid drain-back system are simpler than any other system. Many of the principles drawn from the liquid example will apply to both air and passive systems.

The basic difference among the three types seems to the author to reside in the storage system. A collector is a collector, whether it consists of liquid-filled tubing, air-filled chambers or the wall of a building. All the basic parameters in an active liquid-filled collector find their counterparts in air and passive systems, but the storage is radically different. Compared with a water tank, a rock bin is quite complex; compared to both, the transactions among the masses in a passive design are almost impossible to model exactly. Luckily, in all three cases, simplifications can be found which take advantage of the large averaging effects (i.e., long time constants) of these high-capacitance storage devices.

In a rock store, air exchanges heat with rocks. Immediately, we are faced with a basic difference from water stores: there is a heat exchange process both going into and coming out of storage, so that the store is also the load heat exchanger. Furthermore, while a liquid store can with some slight loss of accuracy be considered homogeneous, rock stores exhibit a pronounced thermal topology which cannot be ignored. The usual procedure is to model a rock store as if it were a series of layers, studying the energy transactions between layers as heated air flows into and out of the store.

The key to taking advantage of the stratification in a rock bed is to arrange the system so that the wave of heat passing through the bed never reaches the end which feeds the collector, at least during cold weather, when the system has no excess capacity. This can be accomplished by reversing the flow of air when discharging the bed (Fig. 7.22), and by making the bed with as long an air flow path as practical, limited

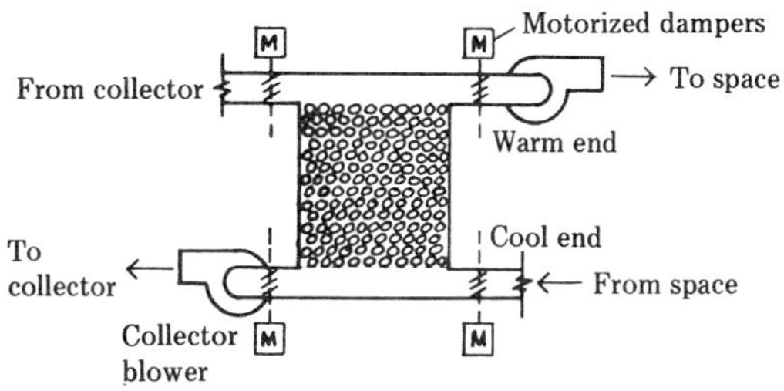

Fig. 7.22 Diagram of a rock bin with reversing dampers, using two fans. Other designs use a single fan, with more complex damping. Two fans are ideal, because each can be suited to its own load.

by the fan energy necessary to force the air through the bed. To see how this works, assume that there is no low limit on the discharge temperature from the bed. As soon as the building needs heat, the fan will force cold return air from the space into the collector-feed end of the store. This air will heat up as it passes over progressively warmer rocks toward the middle of the bin, and then give up some of this heat at the collector-return end, after which it passes through the supply duct system into the space to be heated. This process continues throughout the non-collection hours, until the entire bed, or at least a large portion of it, has approached the building return air temperature. Now, when the collectors next turn on, air at the building return air temperature is fed from the rock bed to the collectors throughout the collection period. Because of the low temperatures of the intake air, the collector operates efficiently despite its low F_R (see Chapter 4).

Why do rock bins behave so differently from water tanks in ordinary solar systems? It is because the storage medium is *static*, in the sense that heat cannot flow from one part of it to the other if the various fans are off. This is in contrast to a water tank, where the storage medium (water) can mix freely until the warmest water is at the top and the coolest is at the bottom. In the discharge cycle, water tanks again behave a little differently from rock bins. Assuming that the water heats an air stream, it is likely that the heat capacity rate of the air will be lower than that of the water. Thus, in a pass through the duct coil, the water would not drop to return air temperature, even if the coil were 100% effective (Fig. 7.23). If the return from the coil were brought through an immersion coil in the tank, the return water would heat the bottom of the tank. For this reason the use of an immersion coil on the load side probably makes little or no difference. The tank would tend to unstratify if the load coil were on. Therefore, it would be better to discharge the return near the top of the tank. At the end of the nighttime

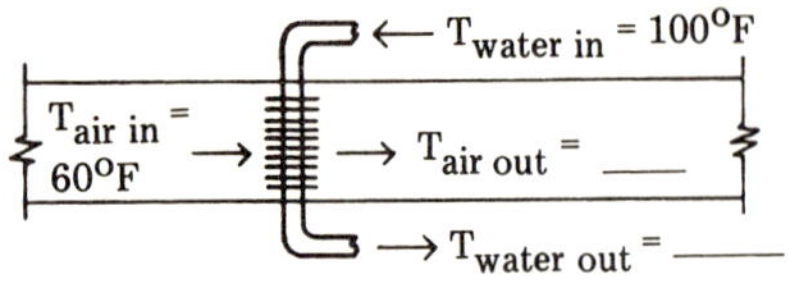

$$\dot{C}_{max} = 8 \text{ gpm} \times 500 \text{ Btu/}^{\circ}\text{F} \times \text{hr} \times \text{gpm} = 4000 \text{ Btu/hr} \times {}^{\circ}\text{F}$$

$$\dot{C}_{min} = 926 \text{ CFM} \times 1.08 \text{ Btu/CFM} \times {}^{\circ}\text{F} \times \text{hr} = 1000 \text{ Btu/hr} \times {}^{\circ}\text{F}$$

$$e = 0.74 = \frac{T_{air\ out} - T_{air\ in}}{T_{water\ in} - T_{air\ in}} \quad (\text{since air has } \dot{C}_{min})$$

$$= \frac{T_{air\ out} - 60}{40}$$

$$T_{air\ out} = 89.5^{\circ}\text{F}$$

$$T_{water\ out} = T_{water\ in} - (89.5 - 60)\frac{1000}{4000} = 92.6^{\circ}\text{F}$$

Fig. 7.23 The drop in water temperature through a good coil is usually a good deal lower than the rise in air temperature, in a well-designed system. This causes the return water to unstratify the storage tank.

discharge period, the entire tank would have approached return air temperature. Here is a major, irreconcilable difference between water and rock storage: Since the water flow rate into a load coil must carry substantially more heat than the air flow, the water can never reach building return air temperature, except after a long period of discharge; whereas, in a rock store, the air is perforce at return air temperature when it enters the store, thereby driving the store down to that temperature promptly, in a wave which travels through the store during the heating cycle.

Controlling an air collector is similar to controlling a liquid-cooled one. The sensor in the rock bin must be mounted to sense the temperature of the air feeding the collector; and because of the low F_R of the collector, the sensor at the collector does not need to be mounted to a detached portion of the absorber (see Chapter 4). On the load side, however, control is greatly complicated by the behavior of the rock bin. At the end of a full clear day of collection, the rock bin will have a temperature gradient which reaches a maximum at the center, and will be at roughly the same temperature at each end—a temperature which could be as low as the space return air temperature. If a low limit is placed on the discharge temperature, where should the sensor be placed? If placed at the end, it will register a temperature below any low limit, and will not allow the bed to be discharged. It must, obviously, be placed within the bed in order to register the presence of "buried" but available heat. After several hours of operation, the space return air will have driven the temperature of part of the store down below the low

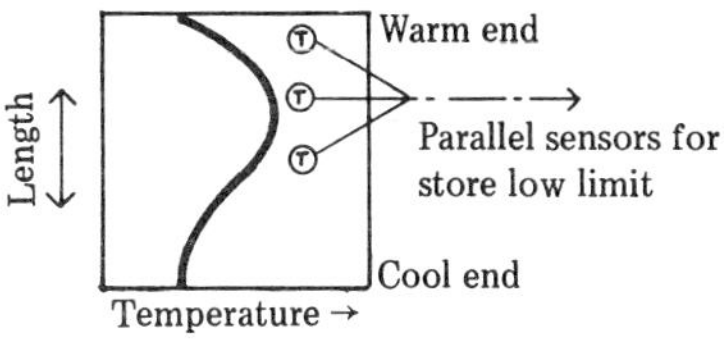

Fig. 7.24 Since rock bins tend to end the day with a maximum temperature buried somewhere within, this suggested sensor arrangement should avoid shutting off the store before it is fully discharged. (The author has not used the scheme.)

limit. If the low-limit sensor is, for instance, dead center in the bed, a large part of the available heat will not be delivered at the time when the sensor shuts the system down (Fig. 7.24).

Two remedies are available, and both are recommended. First, install a series of sensors wired in parallel, so that any one of them can keep the system running, and space them from the discharge end into the middle of the bed. Second, arrange the controls so that the return air always runs through the bed. After the low limit has cut out the operation of the first stage of the thermostat, the second stage will bring on the fan and the furnace. If the fan were always to blow through the bed, what heat is left would be added to the air stream prior to its reaching the furnace, and the bed would often be emptied of useful heat.

Because air flows in opposite directions through the store during charge and discharge cycles, when there is a demand for heat during collection the heat must come directly from the collector. This is a nuisance in designing the control system and duct work, but an inevitable one. With laudable salesmanship, air-system proponents treat this characteristic as an advantage, causing some to wonder whether liquid systems should behave the same way. One is reminded of the company, trying to sell white salmon to a market accustomed to pink salmon, which advertised that *their* product didn't turn pink in the can.

Air systems leak; they require large and cumbersome ductwork, expensive rock stores, and gasketed dampers; they operate at much higher temperatures than comparable water systems; air collectors are surprisingly more expensive in many cases than water collectors, because they contain more metal; and the systems are much more difficult to control. One might wonder why anyone would bother with them. But after you have experienced your first leaky water tank, or ruptured collector tubing, or drain-back pipe which drains the wrong way and causes the system to freeze, then you understand why some people bother with air systems.

LIVING IN A COLLECTOR

Passive solar heating is an attractive idea, because it focuses attention upon the design of the building rather than upon the design of an extraneous artifact to be attached, somehow, to a building which never

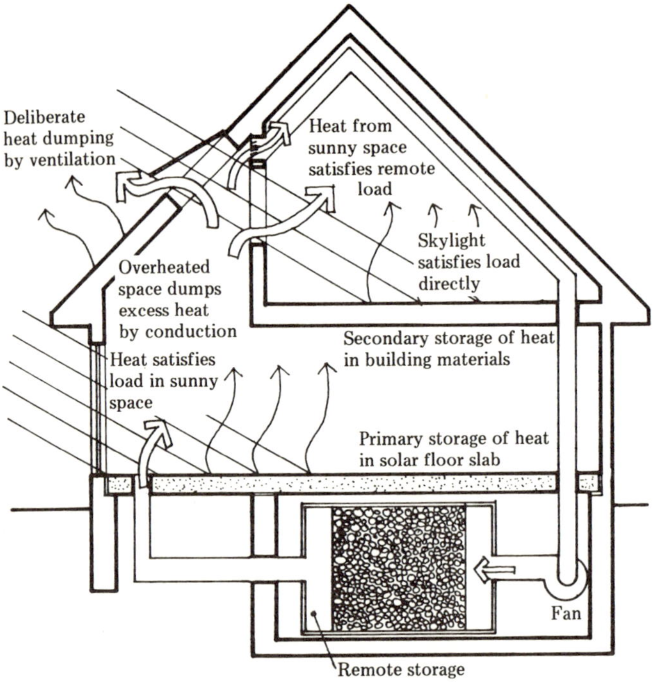

Fig. 7.25 Diagram showing the various destinations for incoming direct solar gain in a residence.

expected such a thing. The author, like many architects, is designing buildings which take advantage of solar gain to satisfy all or part of the load. Using one definition, few of these buildings are passive, because they use some form of mechanical energy to transfer heat. It is best to clarify right away that in this discussion the use of mechanical energy is allowed with passive systems. "Passive" in this context is inaccurate; "intrinsic" might be a better term.

At least four kinds of passive designs have grown up in the last few years: thermosyphoning, direct-gain, greenhouse, and indirect-gain systems. The last three in one way or another are focused on the problem of coping with the sun-pulse. The indirect-gain system interposes a mass between the glazing and the occupant, soaking up the radiation and releasing it automatically after a phase-lag. Direct-gain systems, which are the main subject of this section, use mass in the building as capacitance. The greenhouse behaves like a direct-gain system, but allows greater temperature extremes.

Since passive solar heating is intimately bound up with building design, it is not unreasonable for the design to limit the options available. An argument can be made (one which the author finds convincing) that in most cases it is better to look out of a large south window than to look at the inside surface of a concrete wall. Of course, excellent examples of indirect-gain systems have been built—in particular, the wonderful house by Steve Baer, which allows south light to filter in between oil drums set on their sides. This design, however, probably doesn't fit the image of "house" present in most buyers' minds. In nonresidential buildings, indirect-gain systems may be more appealing, and good new designs are appearing occasionally.

Greenhouses are largely restricted to residences but have great appeal. They are familiar, desirable, elegant, easy to build, and useful. If they are small they can discharge heat to the house during the day, and receive enough heat from the house at night to keep the plants happy. Shades can be used to cut down heat loss. Large greenhouses are better with their own storage; or they can be treated exactly like another room in the house, with direct solar gain. A growing literature describes them and their construction. Their use in helping the homeowner to achieve self-sufficiency has made them popular in a time of rising prices.

Thermosyphoning systems with separate collectors look superficially like active systems. Their operation is contingent upon the development of a substantial thermal gradient between storage and collector, to drive the flow of working fluid through the collector. Aside from a variable F_R, the need for great care in sizing fluid passages, and the need to locate storage above the collector, they can be considered as a form of active collector. The reader is referred elsewhere for further details concerning these excellent devices.

By far the most common and interesting form of passive heating is that involving solar gain through the building's glazing into the living space: "living in the collector," hence the title of this section. It is not a new idea when used in moderation; nearly every building experiences solar gain at some time or other. What is new is the attempt to serve a building's space-heating load largely by the use of solar gain through glazing.

As with any solar heating system, the primary problem is to absorb sun-pulses without overheating the collector. Since the collector is inhabited, the limits to overheating are severe. These limits in turn interfere with heat transfer processes. Not only are many of the available heat transfer operations ineffective, but the limits on overheating prevent the establishment of high temperature gradients adequate to drive the ineffective heat transfer operations. The system lacks both belt and suspenders.

There are five ways of coping with the incoming energy:

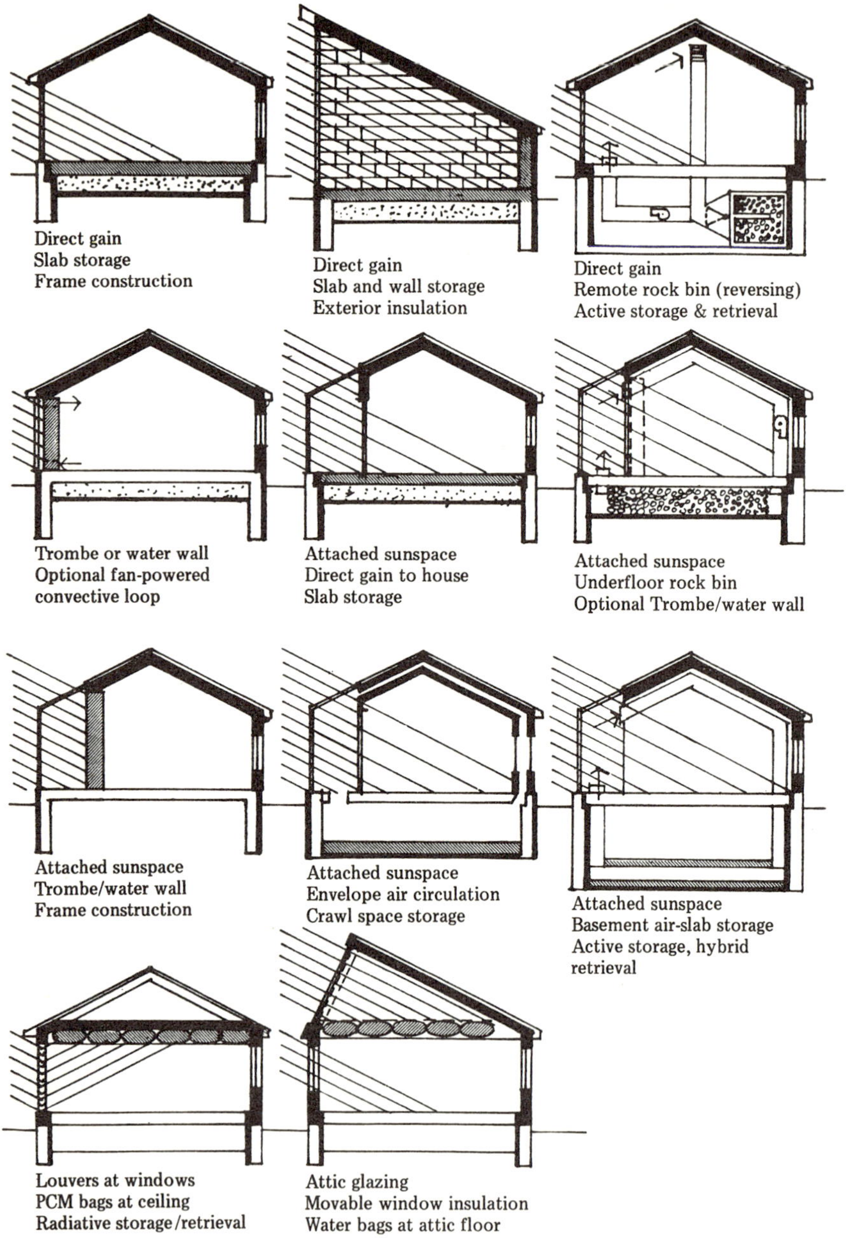

Fig. 7.26 Schematic sections showing 14 different configurations of passive houses.

1. *Use the energy to defray losses.* Here a question of finding a load arises. A passive house is almost by definition highly energy-conserving. The occupant-generated heat can often satisfy much of the demand during the sun period. Losses are reduced because the entire south wall and roof are warmed by the sun—in fact, there is often a small net gain. To maximize the direct use of the sun, it seems desirable either to locate all the rooms facing south, each with a proportion of the solar gain; or to ensure the distribution of air from the south rooms into the north rooms. The latter scheme is usually complicated mechanically, while the former causes severe restrictions to planning.

2. *Dump the energy by overheating.* The easiest way to waste energy is to allow the space to overheat, thereby increasing the loss rate. Since this happens of necessity if measures are not taken to cool the space, it is sometimes hard to recognize the excess heat loss. But losses due to overheating are exactly analogous to the operating losses of an active collector. Counting overheating losses as part of the building load is a classic way of exaggerating the solar contribution of a passive collector, and should be watched for in reports of system performance. Passive design does not need to be justified by legerdemain.

3. *Dump the energy by ventilation.* If energy is not needed, as in the fall, when the sun is low enough in the sky to provide substantial gain, forced or gravity ventilation may be required to control space temperature. The concept again is to increase the demand. Obviously ventilation doesn't work when it is hot outside, at which times the glazing must be shaded. Summer shading is relatively easy on the south, but it is very hard to achieve on east and west exposures.

4. *Store the energy within the space.* In any building, there exists storage within the space simply because (despite years of working at it) we have not found a way to make something out of nothing. Test measurements show clearly the effects of storage in building construction well into the evening, even in a house with wood floors and thin gypsum board walls (see Chapter 6). How much storage to include is likely to be a sticking point for some time. If passive systems are treated as engineering abstractions, intrinsic storage can be required up to 90 Btu/°F $\times$ ft^2 collector, about 3 times the amount used in active systems. However, for years buildings have been evolving away from mass. In single-family houses, every stick of wood that is not essential to the fabric, and maybe a few that are, has been removed in the continuing attack on building costs. A low-mass structure can respond rapidly to changing ambient conditions, thermostat setback, and the converse, morning heating up. Basically, you have a choice in a low-mass building, and can control your climate easily and quickly.

Granted that a minimum of mass ought to be used in a direct-gain passive design, where is the best place to put it? It is well established that mass which is directly irradiated by the sun is far more effective as energy storage than mass irradiated by other warm surfaces in the room. The ideal form of storage is an exposed, dark-colored concrete or stone floor, at least 4″ thick and preferably 6″ or 8″ thick, insulated from moist earth or cold spaces below. Covering this floor with brick slightly reduces its effectiveness; resilient tile causes a much greater drop off; while wood or carpet virtually wipes out the advantage of the mass. You can perform an experiment which demonstrates the effectiveness of various materials as heat sinks. Find a variety of surfaces—rug, wood, resilient tile on concrete, brick, and raw concrete. Stand on these surfaces when they are at 50° or 60°F while in your bare feet. The rug will instantly feel warm, because it is an excellent insulator, and traps the heat from the blood circulating through the tissues in the bottom of your foot. Conversely, the raw concrete will soak up heat from your blood until your foot gets cold and shuts down the blood supply. In between are the other materials, brick the coldest, then resilient tile, then wood. When irradiated by the sun, these materials behave roughly in the same way. The best sinks pick up a lot of the heat and send it into the material relatively rapidly. The best insulators immediately get very warm, and radiate almost all the incoming energy back at the other surfaces in the space, as well as heating the air in contact with the floor and causing convective heat-flow into the air in the space.

The color of the surface struck by sunlight is important—it must be relatively dark to prevent excessive reflection, and to allow the solar radiation to enter the surface of the material and be converted to long wavelength heat energy. Once in the form of heat, however, the energy "sees" almost every surface in a space as dark grey. Shiny metal will be seen as reflective; plastic will be seen as grey (i.e. it is "translucent" to heat); but most finish materials will look almost black, regardless of color and texture. Except for the surfaces which receive direct sunlight, the rest of the room can be light-colored, neglecting the slight advantage of dark surfaces in absorbing diffuse or reflected radiation coming from directions other than that of the direct beam of sunlight.

In deciding whether the next increment of masonry floor or wall is cost effective, it is essential to examine how long the surface is actually struck by the direct beam of sunlight. The strip of flooring right next to the window is the best buy, with decreasing cost effectiveness for each successive strip in the typical case where the side walls begin to cast shadows onto the floor during part of the day. The walls are usually further down the list, unless they are close to glass. The Trombe wall can be seen as a natural response to maximizing the exposure of the storage

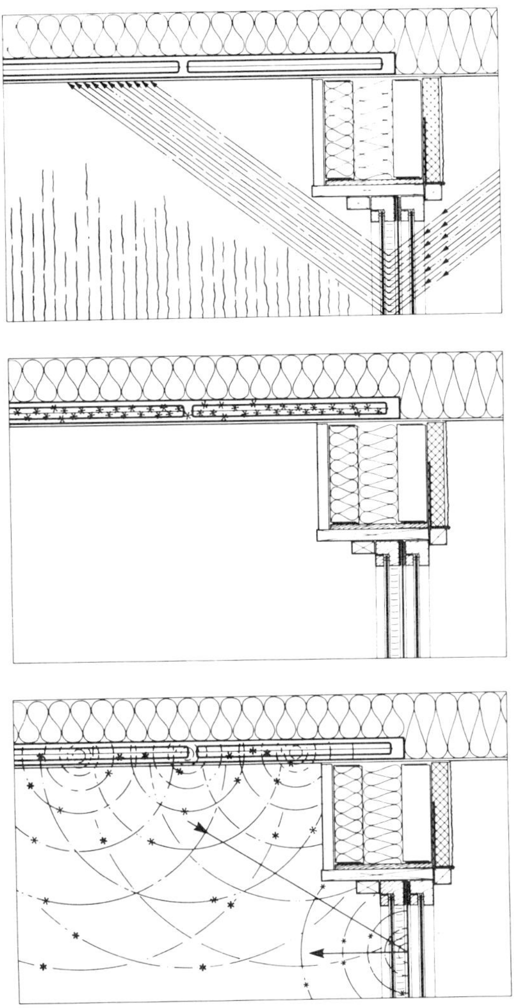

Room heat rises while solar heat energy is directed toward the ceiling by reflective louvers

Heat energy is retained by ceiling panels containing phase-change material

Heat is re-radiated by ceiling panels and prevented from escaping by a heat mirror

Fig. 7.27 Schematic sections showing the operation of the M.I.T. "Solar Five" building. (Courtesy of Timothy Johnson.)

mass to sunlight throughout the day. Unfortunately, one needs a floor in any case, and usually doesn't need a foot-thick concrete wall on the southern exposure.

The concepts introduced at the MIT solar V building, where sunlight is directed at heat-of-fusion mass in the ceiling, deserves attention (Fig. 7.27). The ceiling is usually no place for storage; but this scheme, if it proves out, could make ceiling storage practical, and allow the economical addition of intrinsic storage. The heat-of-fusion concept also

Fig. 7.28 Three-bedroom passive house for the Acorn Structures, Inc. line of homes, designed by the author and his staff in collaboration with the Acorn staff, under a grant from the Solar Energy Research Institute.

helps control overheating, since until the salts are completely melted the temperature of storage stays more or less constant.

Barring heat-of-fusion storage, the temperatures at the surface of the mass are bound to go up as the sun-pulse continues. Ambient temperature also goes up to a broad peak in the early afternoon. For the same reason the available storage in the environment has become saturated with heat at lower temperatures, and higher temperatures are needed to drive more heat in.

One way around the difficulty of providing enough mass within the space is to reduce drastically the heat loss of the building. If energy conservation is applied to very high levels (with commensurately longer paybacks), the residual load in the space can be made so low that a moderate amount of passive solar heating can satisfy a high percentage of the demand. It is very difficult, however, to reduce the gross heat loss rate in a residence below about 4 Btu per degree-day per square foot of house. (Mass-produced builder's houses are lucky to achieve 8 Btu/degree day × square foot.) A 1500-square-foot house in a 6000 degree-day climate therefore will have a gross loss of around 36 million Btu annually, assuming the average house temperature is 65°F. While only a little more than twice the usual hot water load, it is nevertheless a real load. It is easy to see, however, that if 5 to 10 MBtu are available from the heat of occupancy, that a few hundred feet of south glazing could satisfy nearly the entire remainder, given a moderate amount of storage.

5. *Store the energy remotely.* If a south-facing room (or an attached greenhouse) is treated as an *active* air collector with ordinary fans and ductwork to provide heat transfer between the space and a remote store, all sorts of possibilities arise, along with many problems. First, the air flow through the collector must be quite high if any substantial amount of heat is to be removed. Second, even with a high air flow, the equilibrium temperature of the space may be high, due to the very poor transfer of heat between the warm surfaces and the slowly-moving air. One usually installs some intrinsic storage as a first line of defense, removing the remaining heat (if economically feasible) by circulating air through the house ductwork to some kind of remote storage. Rock beds under the floor are popular, but they have the usual disadvantages of rock beds, and since they work at relatively low temperatures, are more likely to mildew under certain conditions (a warm humid day after a cool night). Because they can be nearly 100% effective as heat-transfer devices, they are thermodynamically better than a water-tank and duct-coil combination, which can be at best about 75% effective. Their crucial advantage is that heat can be released from storage passively directly into the space above; a rock bin requires a fan and ductwork.

Living in the collector works only in certain building types. Residences are ideal, since they have continuous loads and are able to tolerate some overheating. Many public buildings are much more in need of *light* than of heat. Successful low-energy office buildings are often based upon the aggressive rejection of solar gain, since overheating is the norm. Some designs employ clever "light scoops" at the outside walls. The MIT Solar V building deflects light to the ceiling with louvers, and thereby avoids glare. In general, passive and active heating must compete with various heat scavenging devices in the nonresidential marketplace. And finally, there can be severe control problems in certain building types if window gain is encouraged.

A BRIEF NOTE ON HEAT PUMPS

One of the most fascinating, as yet unwritten, documents which might emerge from the solar industry would be a compendium of the justifications for various designs combining solar collectors with heat pumps. Most such designs do not improve the performance either of the solar system or of the heat pump beyond a few percent, and the author is perhaps excusably cautious about the potential for these arrangements to work well, when compared with a well-designed solar system working alone. It is, of course, easy to make a solar-assisted heat pump design work better than a poorly designed solar system. There are a growing number of competently made systems combining solar collectors with one or two storage tanks and a heat pump, some of which may well prove to be both cost-effective and maintenance-free. But until this is demonstrated convincingly, the author recommends that the designer follow the precepts of investigators from the University of Wisconsin,* and use heat pumps as back-up devices.

It is important to favor the solar system over the heat pump, since the solar has a much better C.O.P. This is done by placing the solar coil in the airstream before the heat pump coil, turning the solar on first, and only running the heat pump if the solar system can't handle the load. If the solar coil is placed *after* the heat pump coil, it will have to operate in an "either/or" rather than "both/and" mode (just as with a hydronic backup); or else a low limit on the solar system must be set at the output temperature of the heat pump, which might thwart the use of low-cost, high-loss collectors. Unfortunately, most packaged heat exchangers place the coil in the air stream before the fan. If the solar is before both, the fan and possibly the coil may be overheated when the solar tank is

---

*Mitchell, Freeman and Beckman, "Heat Pumps: Do They Make Economic and Performance Sense with Solar?" *Solar Age*, July 1978, pp. 24–28.

Fig. 7.29 The interior of Massdesign's prize-winning Thompson House. Living in the collector can be a delightful experience.

warm (in spring and fall). A heat pump made of separate, modular elements solves this problem.

At this writing, goundwater-source heat pumps are not common, with the few well-designed units available priced in the luxury range. Briefly, watch out for: scale from minerals in the groundwater; the need for very high flow rates to compensate for the low source temperatures; low C.O.P.s of the units, again due to the low temperatures; the pumping energy cost out of and back into the ground; the cost of drilling two wells, and possible problems with the local environmental authorities if a significant amount of heat is added to the groundwater in the cooling mode (if any). Most ordinary water-to-air heat pumps, designed for use with a run-around heat pump system, are not applicable to well-source systems, unless the groundwater is quite warm.

Observing these caveats, which really are only an indication of the early stage of development of these devices, by all means investigate this promising area.

8

WATER HEATING

SUPPLY, DEMAND, AND STORAGE

Plotting the solar supply and water heating demand curves for a typical climate, as in Fig. 8.1, it is easy to see that water heating is on a monthly average basis much more "load following" than space heating; that is, the supply and demand curves match more closely. On a sunny day, the supply and demand again are closer than in the case of space heating, since much of the water is drawn during daylight hours. Of course, if the weather is cloudy, the demand and supply are totally mismatched in any solar system.

As opposed to space heating, which we need (with some variation) all the time during cold weather, water heating is an adjustable demand. Indeed, in many simple solar water heaters, demand is contingent upon supply—no sun, no hot water.

This flexibility in demand is useful in understanding the nature of storage in water heating. Suppose instead of a storage tank, we were to install a truly instantaneous heater, as is frequently done in Europe. If the heater is an electric resistance element at the faucet, it will—if economically sized—heat only a small stream of water to the desired use temperature. The flow rate is determined by the water's incoming temperature, its outgoing temperature, the effectiveness of the heat transfer device, and the temperature of the heating element. As the incoming temperature rises, or the outgoing temperature falls, the flow rate can theoretically go up. In the U.S., an intermediate design is often used, namely a heating coil immersed in a boiler used otherwise for space heating. The unit can handle a stream of water the flow rate of which depends upon the firing rate of the boiler, the size of the coil, the storage capacity of the boiler and the maintained temperature of the boiler water, as well as the incoming and outgoing temperatures of the water stream. The behavior of such designs is very complicated, but universally they suffer from one defect: they require the boiler to be kept hot throughout the year. The resulting skin loss from the boiler and losses up the stack can reduce the unit's efficiency to very low levels. It is possible to overcome this defect to some extent by adding a storage tank to the system, firing the boiler only occasionally, and pumping water from the tank to the boiler during firing; many installers will add such a unit to an existing boiler.

But in the U.S., the basic water heating system involves a storage tank with an integral electric resistance element or gas-fired heat exchanger. The storage is there simply to match a variable *demand* to an undersized *supply*. The tank, by analogy with a solar system storage tank (which matches a variable *supply* to a continuous or variable *demand*) also provides capacitance to deal with sharp peaks—in this case *demand* peaks instead of *supply* peaks. As in a solar system, a trade-off is made between the size of storage and the size of the supply unit. The larger the unit, the smaller the supply unit can be, up to the obvious limit of a heater running nearly continuously. There is usually little penalty to pay for increasing the heating unit size, so that in practice standard designs are made, with the heating rate rising along with storage capacity, as larger units deal with progressively larger loads.

Storage is useful in allowing a relatively small, central heating device to serve a large load. In addition, it provides capacitance to deal with sharp pulses of demand, in which a lot of hot water is drawn in a short time. Instantaneous heaters are generally uneconomical if designed to serve such sharp peaks, and (except for multiple units of demand, where all the units do not draw at the same time) instantaneous heaters set a top limit on the rate of draw. Storage is a convenience which allows us to be very casual about when we draw hot water.

When solar is added to a water heating system, a new kind of storage is introduced. Its function, as noted above, is the exact opposite to that of hot water storage. The two storage volumes should not be mixed, since we want the collector to heat the coldest water in the system, and raise it only to the minimum temperature required to "use up" the available solar energy. Suppose, for instance, that on the average in midwinter, the solar system could supply only about 50% of the hot water required. It would be better to raise *all* the water to an intermediate temperature than to raise only some of the water to use temperature, since the collector would in the first instance operate at a lower temperature and deliver more energy. More exactly, it would be better to deliver 55% of the energy at a low temperature than 45% at the use temperature (percentages arbitrary). As usual, this requirement relaxes when very low-loss collectors are used, or when the climate is very mild and collector operating temperature is not as severe a limitation as it is in cold areas.

The sizing of storage also responds to solar needs, and not to the need for hot water. Typically, one would like enough storage to keep the collectors operating at low temperatures. If the supply is large relative to the demand, the large storage will rise in temperature, indicating an energy surplus. While some of this will be used to satisfy the larger skin losses from the larger tank, on the average the larger tank will save more than it loses. As in space heating systems, the limit is set by the cost of the tank and its installation, coupled with the cost of controlling the skin losses through thicker insulation. Many early (1975 era) solar water heaters failed because of excessively small tanks.

In many utilities, off-peak rates are available for water heating. Two elements are often used: the one on top for emergency on-peak heating, with the one on the bottom for off-peak heating. In effect, two storage volumes are present—a small volume at the top for immediate use, and a large one at the bottom for storing energy. Exactly the same principle is used in most solar water heaters. Solar is used to heat a large volume at the bottom, with the auxiliary energy heating a small volume at the top. If the solar is supplying more energy than is consumed, the temperature of the bottom volume will rise above that of the upper volume, shutting off the aquastat controlling the auxiliary heater. These designs are possible because of the tendency of water, if introduced into the tank carefully, to stratify very strongly. In one test, the boundary layer between an upper volume at 140°F and a lower one at 40°F was only 4″ thick. Indeed, some standard water heater designs include devices to upset purposely the stratification in order to prevent very high temperatures from building up at the top. If such devices are not used, the top few inches of storage are wasted, because the high temperatures prevent that water from being used.

In general, unless a low-loss collector is used in a mild climate, the storage volumes for ordinary use, for off-peak use, and for solar use must be kept separate. And, neglecting cost, such separation will always enhance solar output if skin losses are kept under control.

ENERGY CONSERVATION AND WATER HEATING

As in any kind of solar installation, energy conservation should come first. Paradoxically, a recent development, if successful and cost effective, could do such a good job of energy conservation that solar water heaters might become a thing of the past in many climates. Under test or on the market are water heaters using a small water-to-air heat pump. The device works at a C.O.P. of 2 or 3, in the latter case reducing the electric consumption of the water heater by two-thirds whenever the heat pump can run. In warm climates, the byproduct of the water heater, namely cold air, is extremely useful. One can therefore write off another two-thirds of the electric consumption against air conditioning (provided, of course, that one needs air conditioning; this point may be overlooked in the promotion of the device). So, in the air conditioning season, the device has a net consumption four-thirds less than a normal resistance heater! If the air conditioning season is short or absent, a saving of two-thirds is possible as long as the building can stand being used as a heat source. Sooner or later the basement or other space where the heater is located gets too cold, and normal resistance heating must be used. Unfortunately, this will tend to occur in cold weather, when solar energy is at its lowest supply. As usual, heat pumps and solar are in competition, both "crashing" at about the same time.

One very interesting speculative use for a heat-pump water heater would be in conjunction with passive solar heating. If water is heated primarily during collection hours, the cold discharge from the water heater could be used to hold down operating temperatures in a solar space, thus extending the useful collection hours and increasing the yield. It is important to remember, however, that you can't in this case count the energy twice, as you can in the air conditioning case. Either you credit the heat pump with running longer at the expense of a larger heat source, or you credit the passive collector with extended operation at the expense of the electricity used to run the heat pump. (The extra heat collected is used for water heating and not for space heating.)

Two other devices resemble the heat-pump water heater. The first, available since 1979, is a piece of equipment added to an ordinary heat pump which allows the heat normally rejected to the environment during the cooling cycle to be used to generate hot water. The second, for which no workable designs are known to the author, is a combination of a refrigerator and water heater, with the heat rejected by the refrigerator used to heat water. A back-up heater would be necessary, but the device should work well, at least in theory. It can be argued that heat rejected from a refrigerator is useful in the winter, but the counter-argument can be made that it is rejected into the one space which needs it least, the already-overheated kitchen. Much of the potential heat of occupancy attributable to the refrigerator must be discounted as causing overheating, especially in very energy conserving houses. Using this heat to generate hot water seems very sensible, and there seems to be a close match between the energy rejected and that needed for water heating. No doubt such a device is in the works somewhere. How many other hidden synergies are there in our energy consumption patterns? It is exciting to imagine that there are many more.

Given ordinary water heaters, the first energy conservation measures are insulation on the piping and extra insulation around the tank itself. Getting a plumber to insulate water piping is like pulling teeth, even with today's energy prices, but it is a must. Extra insulation around the tank, if simple fiberglass batting, is usually cost effective.

Next, demand can be modified. If off-peak rates are available, they are usually very worthwhile, providing they apply only to hot water. If one must pay extra for the remaining electricity while paying less for off-peak water heating, it is probably not a good buy in most cases (for residences). Flow rates in showers can be reduced, and in general, less water used (for instance, by running the clothes and dish washers on a shorter single-wash cycle).

Perhaps the most questionable measure is the reduction in use temperature. Fig. 8.1 shows that, *if the tank must be made larger* as a result of reducing the use temperature, one gets very little thermal benefit, and may lose with regard to cost effectiveness. On the other hand, if use temperature can be reduced without disturbing the desired consumption patterns of the building, it is always a good idea. Sometimes, if electric-resistance heating is used for water heating, electric-resistance boosters can be used in the clothes washers and dishwashers, which require much hotter water than ordinary washing, thus allowing the tank temperature to be reduced. Better still, wash clothes and dishes at lower temperatures.

In large buildings, very long piping runs from the heater to the faucet require that hot water be circulated at all times through the water piping by means of a pump actuated by an aquastat. This procedure, even with well-insulated piping, is very wasteful of energy, since it in effect extends the walls of the water heater out into the piping system and greatly increases the skin losses of the system. Several measures are possible, such as point-of-use heaters, distributed water heaters, etc., but they all suffer from high maintenance costs. The most energy-conserving measure is to eliminate the recirculating system, but users would find this intolerable, and water consumption would go up, increasing an already serious problem in most localities. In the case of recirculating loops, a reduction in water-heater temperature is a very good strategy, since it reduces skin losses throughout the piping. Remotely-located clothes washers and dishwashers would then be fitted with booster heaters.

Probably the most effective energy-conservation measure in larger buildings with rental units is individual metering. This excellent measure unfortunately removes any incentive for solar heating, since

Suppose 30 gallons of water at 100°F were needed in a continuous stream. Incoming cold water is at 60°F, while the stored hot water is at 140°F. To supply the load, 15 gallons of cold and 15 gallons of hot water are mixed at the faucet. The water tank must be sized at 40 gallons to allow further draw while recovering from the demand.

If the tank stored water at 100°F, it would have to be twice as large, since the entire flow would come from the tank. Calculating skin losses from tanks:

	40 Gal.	80 Gal.	Difference	Value of Saving (@ \$10.00/MBtu)
Diameter	16″	20.5″	—	
Height	46″	56″	—	
Insulation	R15	R15	—	
Temperature	140°F	100°F	—	
24-hr. heat loss				
room at 50°	2736 Btu/day	2400 Btu/day	336 Btu/day	\$1.23/yr.
room at 55°	2584	2160	424	\$1.55/yr.
room at 60°	2432	1920	512	\$1.87/yr.

The example shows that a trivial saving in skin loss occurs. If the boosters are electric and the main water heater is gas, expensive energy is substituted for cheap. A more expensive tank is used.

Fig. 8.1 Comparison of domestic water stored at high and low temperature. A crucial assumption is that neither tank is oversized.

the tenants usually cannot get together to install such a device (especially on a building they don't own), and the owner is unlikely to add capital cost and the resulting higher rents in order to save the tenants money on their energy bills. Condominium associations provide a medium for the addition of solar heating in that increasingly-popular form of ownership, thus overcoming some of the resistance to solar in multiunit buildings with separate metering.

If solar heating is present, the occupants can improve the solar fraction by drawing hot water during collection hours. This strategy insures that the solar collectors will be working with the coolest possible water during operation. A microprocessor coupled with delayed washing cycles could accomplish this measure. Notice that it is necessary to move consumption from evening postcollection hours *into* the collection period in order to accomplish this saving; using water just before collection is the worst possible strategy, since it guarantees the highest possible tank temperature during the night, and accomplishes no drop in collection temperature (since the daily consumption is the same and the storage, therefore, enters the day at the same temperature regardless of when the water is drawn).

In a building requiring hot water at 180°F (most restaurants), it might be worthwhile to add a separate heater operating at that elevated temperature, with the basic heater at a lower temperature. If solar were present, it would preheat all the water; if enough solar energy were available to do all the heating, the solar storage would be above 180°F and both heaters would be idle (naturally, a tempering valve should be installed in *all* solar water heater applications downstream from the auxiliary heater).

AVAILABILITY OF DATA

Much remains to be done with regard to measuring hot water consumption. Right at the start, confusion exists because of our habit of measuring the consumption in units of *volume* rather than *energy*. If we say that a family of four uses 75 gallons of hot water per day, do we understand that every gallon has been raised from cold water temperature to the temperature of the water in the heater? or to use temperature? Do we know what the cold water temperature actually is? Do we know how much energy is used to replace skin losses?

If, as in the standard drain-back design used as an example throughout the book, water is preheated in the solar tank by means of a heat exchanger, another unknown enters. How much water moves throughout the heat exchanger (in this case, an immersion coil) at each of the infinitely variable flow rates possible from zero to maximum? If maximum flow rarely occurs, the coil would not have to be sized for such a flow; and if most water was drawn at a particular range of flow rates, we would have a good design goal for sizing the coil. Unfortunately, almost no data is available on the subject. Existing data is either oriented toward the sizing of storage to accommodate demand peaks, which requires only knowledge of peak *hourly* flow rates; or the sizing of piping systems, which requires only knowledge of *maximum* peak rates. No one ever needed to know the distribution of flow rates, and so the data was never generated, as far as the author knows.

The distribution of consumption during the day, on an hour-by-hour basis, has been derived for residences, and is shown in Fig. 8.2. This curve is, however, based upon the averaging of several reports, each of which varies remarkably from the other. While there is uncertainty because of a lack of data for buildings large and small, there is

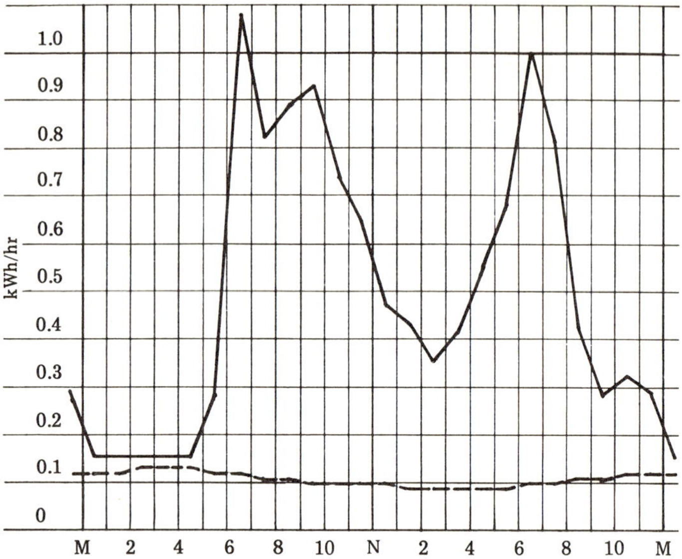

Fig. 8.2 Domestic water consumption for an average family, averaged hourly for a typical day. Vacation consumption, equal to tank stand-by loss, is shown dotted. (See Fig. 6.5 for source.)

intrinsic uncertainty for individual households, even assuming perfect data. This is due to the variability of people's behavior, and to the variations in the pattern of family life over the years as job conditions, children, schooling patterns, and home occupancy change. By contrast, larger buildings (such as apartment houses and office buildings) ought to have a lower unpredictability in hot water consumption. While probably true for luxury apartments, it was certainly not true in the case of two apartment buildings for the elderly in northern New England during 1978, both with solar water heating systems. In one case, the solar system was sized to provide 50% of an expected demand of 2600 gallons per day. The actual demand varied, but averaged between 400 and 500 gallons per day, about 1/6th that expected! Such gross errors are probably more common than is admitted in public. It is urgent that an effort be made—perhaps by ASHRAE—to generate up-to-date data, and keep it up-to-date.

Diversity in demand is to some extent a known quantity, and is helpful in deriving useful flow-rate data. It is notable that different building types have widely varying diversity effects, varying from almost none in motels (all the resident businessmen rise at the same time) to a great deal in some types of public buildings, where people might use the toilets at random times.

SYSTEM TYPES (FOR FREEZING CLIMATES)

There are almost as many system configurations for solar water heat as there are people to think them up. In general, because of the poor heat transfer from air to water through coils, and the danger of coil freeze-up due to air leakage, solar water-heating systems with air-cooled collectors have not been widely adopted. Liquid-cooled systems might be classified as thermosyphoning, drain-down, drain-back, and antifreeze, based upon the heat-transfer method in the collector loop.

- Thermosyphoning systems are simple, reliable, and trouble-free if properly designed. The crucial design points are that the tank be located above the collector and that the piping be sized so that water will rise rapidly enough within the collector loop to produce an acceptable F_R and thereby enhance efficiency. How the water is pumped into and drained from the tank varies widely, and depends upon the likelihood of freezing. A variation on the thermosyphoning system uses freon, which undergoes a phase-change during the collection cycle, and which does not freeze at ordinary temperatures.
- Drain-down systems use the tap water to be heated as the working fluid, draining the collectors whenever the outdoor temperature drops below a preset amount, usually 36°F. Although widely used, these systems rely upon the positive action of an electrical valve (or two) to prevent freeze-up, and have suffered from freeze-up problems. Also, tapwater can be unpredictably corrosive, and therefore a poor working fluid, especially in cities.
- Drain-back systems use repeatedly a fixed volume of tapwater, treated if necessary, which drains back into storage whenever collection ends. Drain-back systems in general form the subject of much of this book. Drain-back water heaters use a preheating coil in the storage tank, usually a coil of copper tubing with or without fins, and may or may not include a recirculating pump to allow the solar tank (which is totally separate from the water heater) to satisfy standby losses when the solar tank is hot enough. Some designs have been developed which use a small holding tank and a heat exchanger between the stored and circulated water, and the tapwater to be heated; but these designs are esoteric and not widely used. Drain-back systems require piping which drains to storage—a difficult restraint for retrofit installations, which form the bread and butter of solar water-heating installations.
- By far the most common system type in cold climates is the antifreeze system. Because a pressurized loop is used, ordinary heating practices can be used, components are easily packaged, and antifreeze piping can wrap around and through the worst existing conditions. Also, various designs of storage tank are available, using wraparound or

immersion-type heat exchangers with auxiliary electric coils at the top. If gas is the back-up, a second tank is typically used for the solar alone.

GUIDELINES FOR DESIGNERS

1. Data on the time distribution of flow, required in order to size instantaneous immersion heaters in a solar tank, are not available. Some data can be found correlating peak flow with daily flow, and coil can then be sized based on some large fraction of this peak flow. In residences, the author guesses that most of the water is drawn at about 2 GPM.

2. For annual load simulations, a rough distribution of the hot water draw during a typical day is needed. In general, residential loads peak at mealtimes.

3. In residences, the draw varies dramatically from user to user, and even over time for the same user, as the family adjusts to the need for energy conservation and to changes in life patterns.

4. The draw rate is a meaningless number unless you also know the temperature rise in the water drawn. Incoming water temperature varies a great deal, and little hard data can be found. In general, city water held in reservoirs ranges widely, while well water stays relatively constant in temperature. Standpipes expose stored water to solar radiation which can raise its temperature in warm weather.

5. In choosing a backup energy source for a solar installation, it is better to stay with the least expensive fuel, unless the solar fraction is above 70% or so, at which point the convenience and higher efficiency of a single tank with electric resistance backup makes electricity a good choice. If a tankless heater in an oil-fired boiler is present in a retrofit installation, try to make the solar system large enough to allow the boiler to be shut down all summer; when the boiler is running anyway to provide space heat, the penalty for using it as a backup source is quite small.

6. If a two-tank system is used, as is usual with natural gas backup, it is possible to add a recirculation loop with a pump, to allow the solar tank to heat the water in the gas-fired water heater if there is no draw. This is necessary only when the solar fraction is high enough that the temperature in the solar tank will normally be above the setting of the auxiliary tank during warm, sunny weather. The loop allows the surplus solar energy to be used to make up for the standby losses in the auxiliary tank when there is no hot water being drawn. Such a loop is probably a good idea in an office building, unnecessary in an apartment house (where some water is drawn every day), and marginal in a resi-

dence, where the owners could shut off the auxiliary if they plan to be gone for any length of time.

7. Unless a very low-loss collector is used, and some clear cost advantage accrues from the design, do not feed the solar collectors from the part of storage heated by the auxiliary. Always try to let the solar system do *part* of the work *all* the time, rather than *all* the work *part* of the time.

8. Energy conservation comes first, even if it removes the need for a solar system. The purpose of solar after all is to save energy, and if other means can accomplish the savings at a lower cost, they ought to be used.

9. If an active liquid-cooled solar space-heating system is present, use it also to preheat water. Only with passive or air collectors is it reasonable to install a separate solar water heater.

9

SYSTEM OPTIMIZING

DEFINING THE OPTIMUM

An optimum system is one which cannot be any better.

Perhaps that is sufficient, and we should go on to the next chapter. Most of a designer's time is spent juggling so many variables that simple adequacy seems like a goal, with optimization a vision of Utopia. It therefore requires a certain daring to explore a subject so remote from the actual achievement of the author, and presumably most readers.

First, a solar-heating system must fulfill the myriad demands of construction. It must be safe, it must be durable, it must require little maintenance. A long list of the issues which must be considered in any building project solar heated or not, can be found by reading the requirements written in building codes, zoning ordinances, and other legal devices. Such considerations lead forcibly to *simple*, replaceable parts, a characteristic of systems which operate at low temperatures.

Second, given a durable set of parts which meet codes and can be built in a maintenance-free manner, how big should the parts be? What loads should they serve? Part of the answer lies in site planning, discussed in Chapter 3. Part lies in the cost of mounting or housing the device, a cost which should always be reflected in the cost assigned to the device itself. For instance, a simple tank can be built to rest in a basement, for 60 cents a gallon, installed, 1979 prices. But how much is the basement space worth? Then again, how much is the heat lost from the tank worth, compared to a tank which loses heat into the ground?

Finally, optimization takes on a technical meaning, when system simulations are used to find the "best" size and configuration for the system and its parts.

System A:	
Cost: Output:	$8,000 30 MBtu replacing oil @ $10/MBtu 10 MBtu replacing electricity @ $20/MBtu
Energy payback: Dollar payback:	8000/40 = $200/MBtu annually 8000/300 + 200 = 16 years

System B:	
Cost: Output:	$8,000 10 MBtu replacing oil @ $10/MBtu 25 MBtu replacing electricity @ $20/MBtu
Energy payback: Dollar payback:	8000/35 = $229/MBtu annually 8000/100 + 500 = 13.3 years

Using an energy criterion, A is 15% better; using a cost criterion, it is 17% worse, when compared to B.

Fig. 9.1 Comparison of energy payback and dollar payback for two systems replacing different energy sources.

Before discussing the simulation procedures, it is crucial to define exactly what is optimum in economic terms. Are we trying to get the most *money* for the dollar invested? or the most *energy*? In the HUD Solar Demonstration Program, the latter measure was adopted in later cycles. It proves to be a good one, although somewhat abstract to present to potential customers. It is the measure we propose to use, but it is only fair to point out that there are circumstances under which it will give the wrong answer.

The measure of cost effectiveness proposed is the dollar value invested per unit of average annual net output. Typically, these are expressed in dollars per million Btu, or per kWh. A good system in New England, for instance, can achieve $200 per million Btu, or $.67 per kWh (1979 dollars). Price the energy and you have a measure of payback. If a million Btu can be obtained in a good system from 11 gallons of fuel oil, and if the oil costs $1.00 a gallon, the simple-minded "payback period" is 200/11 or 18 years.

The "wrong answer" occurs when the energy replaced has two different values, for instance, electric hot-water heat and gas space heat. One could, as in Fig. 9.1, design two systems, one achieving the highest *energy yield*, and the other skewing the output toward hot water to provide a lower energy yield but a higher *dollar return*. We question whether such sophistry ever occurs to a designer; and who knows what the ratio of electricity to gas costs will be in the future? And what happens if you use off-peak energy for the hot water back-up? Such considerations lead us toward the energy yield measure, despite its potential for misinforming us.

PAYBACK

Some of the most impressive by-products of the solar industry are the cost-analysis procedures. Drawing upon sophisticated business planning techniques, they help immensely to simplify the complex trade-offs involved in predicting the outcome of future economic transactions.

Unfortunately, the future is cloudy. As population pressure builds up, with the attendant development, land use, resource, environmental, and energy problems, small events can trigger large consequences. Values change more rapidly; formerly certain conditions no longer prevail. In particular, the crucial ratio between the rise in energy cost and that of the General Inflation is unpredictable. The consequence of this uncertainty is the erosion in confidence of any projections, no matter how arcanely framed in terms of discounts, present value, or internal rate of return.

Given such uncertainty, why has so much effort been put into elaborate economic projections? Simply because solar has proved to be far from cost effective by ordinary measures. It is instructive to ask a developer or other entrepreneur how he measures the worth of a project. While using very sophisticated analysis to fine tune his decisions, he is likely to use some extremely simple test. For instance, he might take the total project cost and measure it against the total expected new annual revenue. If the former is more than 8 to 10 times the latter, the project is dropped.

Solar buyers probably use similar decision tools. The author recently polled the students at the opening of his solar class as to how small an annual saving they would accept from a system costing $2,000. The amounts ranged from $0, reflecting a high moral value placed upon solar (or else a good sense of humor), to $500. The average was close to $200, one tenth of the cost. There is magic in this 10-to-1 ratio. You will hear optimists quoting 7-year paybacks, presumably because they know it will sell, being lower than 10.

The point is simple: payback period, in the ordinary sense of the word—the ratio of the investment to the first-year return on that investment—is a sound and reliable rough measure of the investment's worth. If the consumer feels energy prices will rise rapidly, he will soon enough extend his acceptable payback to account for that. But no one will stand still for the 40- and 50-year paybacks (i.e., 2.5% to 2% return) common with many present-day systems.

There are two ways to shorten the payback period: reduce the size of the investment or increase the first-year return (naturally assuming it is representative of many subsequent years). Tax credits reduce the investment, while relief from property tax, low maintenance costs, and low operating costs increase the return (or more exactly, decrease the erosion of the return caused by such factors). Increasing the cost of energy helps a lot to reduce the payback period, but such increases operate upon both the cost and benefit in unknown proportion.

It is self-evident that energy conservation measures of all kinds should be undertaken whenever possible, up to the same payback period as the solar equipment, before solar is added. If incentives are enacted to reduce the payback period for solar, equivalent incentives are needed to encourage energy conservation at the same levels as before, to prevent the curious situation of less economical measures being used first.

There is, of course, always a percentage of buyers who ask to be coerced into buying a system. It is under these circumstances that more complex analyses are useful—indeed, indispensable. It is often helpful to run a string of calculations taking account of actual out-of-pocket expenses year by year. For a high-bracket taxpayer, the result often indicates a break-even point several years before the end of the "payback period." But many people are disappointed with such analyses, for the same reason they are disappointed with the payback—you just don't get enough for your money.

The author has found that, almost no one will buy "ordinary" (less than 100%) systems with 25-year paybacks; some will buy systems with 10-year paybacks; that most will buy at 5-year paybacks; and nearly all will buy when the payback is 2 years. The resistance to the sales of solar heating is not, as some claim, because it is poorly marketed. It is poorly marketed simply because it is too expensive. If we get the cost-effectiveness up where it belongs, then the "marketing boys" can do their stuff. The designer's job is cut out for him: reduce the payback period!

ESTABLISHING SYSTEM COST

System cost is a surprisingly difficult thing to pin down in many cases. On the one extreme is the retrofit swimming pool or water heating system. An installer will quote an exact price for a complete add-on. No confusion exists, since only one project is involved. At the other extreme is a direct-gain passive design. Defining the strictly solar elements is virtually impossible, since almost any house has solar gain which can be (and usually is) used constructively to reduce heating costs.

In the difficult case, the passive house, it matters whether the house falls in the luxury or the low-cost price range. An owner who can afford special amenities might choose to divert money from one amenity into special construction for purposes of passive gain. If a satisfactory house can be built which, along with most of the required amenities, also has a lot of passive gain, the gain can be considered "free" since the overall project cost has not risen. In the case of an owner who is watching the budget, the cost of nonsolar construction is usually known with reasonable accuracy. Any changes made to accommodate solar heating, whether active or passive, can be traced and examined for cost effectiveness. It is, however, easier to lose the cost of windows and masonry walls than it is the cost of collectors, pumps, and storage tanks.

If the solar construction has a second use, the solar cost can be reduced by the value of that function. Direct-gain passive systems benefit more from this than do Trombe-wall designs, simply because a south window which can be looked through has an architectural function more highly valued by the user than a concrete wall. Steve Baer's design, where light filters between water drums, creates as much or more architectural delight than a large window—especially in summer. But in the author's experience, most people would gladly exchange the best of Trombe- or drum-walls with a wall of clear glass, providing that glare and overheating are kept under control.

Besides other uses, solar materials may have structural or weatherproofing functions. Some collectors replace all or part of the roof, and a credit can be taken for the displaced elements. Massive walls hold up the roof and glass walls replace opaque ones. Backup systems can sometimes be reduced in size when coupled with solar or energy conservation measures, and credit taken.

In both active and passive design, the overall form of the building is sometimes distorted in order to accommodate the collector. The plan of a direct-gain passive house might be somewhat less efficient if all rooms are facing south, and there might be more perimeter than in a normal design. For active systems, the steep roof required places major constraints on the planning. Where a small collector is added to a standard ranch house, a boxed-in "hump" in the roof is often used, with an identifiable cost. Finally, flat roofs with parallel collector arrays require supporting frameworks, whose cost can on occasion equal or exceed the cost of the collectors themselves. (It is usually cost effective to use low-loss collectors in such situations, in order to amortize the cost of the supports over the largest possible output—or, put another way, minimize the size of the array, and therefore the cost of supports, for a given output.)

DETERMINING SYSTEM OUTPUT

It is very difficult to determine the exact output of a solar heating system. Ideally, one would like to know the change in ordinary energy consumption in a building before and after the installation of solar. In practice, this is usually an unrealistic goal. For new construction, there is no "before"; and for retrofits of existing construction, so many other factors are likely to change that the before and after measurements are not comparable (except for some retrofit water heaters and swimming pool heaters).

With elaborate instrumentation (such as that installed in many HUD and DOE demonstration projects, working now after several years of start-up problems), it is possible to track enough of the thermal transactions in a building to plot the system output with some accuracy. Clearly such an approach begins to cope with faulty installation, undetected air leakage, control failures, etc., and as such is useful. Supposedly, out of this data collection activity will come a giant computer model which will tell us all—or tell us so much that we know less than we did when we started. In the end, computer simulations, or nomograms derived from them, will be the basis for determining system performance. Each component can be tested, even the whole system, but the interaction of system and building requires "hands-on" modeling to test the wide variability experienced in real installations. In particular, if a particular design deviates substantially from the norm, only a simulation can predict the system output.

It would be ideal to test simulations against real data, and such tests will become increasingly common as reliable field data accumulate. Lacking such certification, a simulation can still be very useful in exploring the effect of incremental changes in certain key system variables. It is important to be able to change the *building* parameters as well as the system parameters. Simulations such as "F-chart" lack this ability and must be used with caution if applied to new or innovative designs. There is a need for a variety of tools, each fitted to a particular need.

GUIDELINES FOR DESIGNERS: GENERAL PRINCIPLES OF SYSTEM OPTIMIZING

1. If you attempt to optimize a system without looking *outside* the system for potential optima, you probably won't ask crucial questions, and therefore will miss the best answer. For instance, if you aren't looking for seasonal effects, you might not think to model a large store with very good insulation. After all, a simulation or design procedure is itself a model of a preconceived system, and it is not surprising that the optima reached must be those which derive from the preconceptions inherent in the procedure. The government sponsorship of research subtly encourages research contractors passively to plot variable against variable, with never a creative thought about "what would

happen if . . . ?" This book is an attempt, among other things, to stimulate creative thought through the understanding of underlying principles. It is this kind of understanding which is needed before optimized systems can be found.

2. The essential element in optimizing is incremental change. Does this next increment of storage, for instance, increase or decrease the overall system cost effectiveness? Jumping from a bad example to a good one without considering each step between can lead to absurd results. For instance, let us use this wrong procedure to discover the optimum amount of insulation in a piece of wall. Start with a totally-uninsulated wall with an overall R-factor of 5. Assume 6¢ per R per ft^2. If we are satisfied with a 10-year payback, how much insulation is optimum in a climate with 6000 degree-days, with the interior space kept at 65°F all the time, and with energy worth $10.00/MBtu?

Answer: R-42. If, however, we start with a wall of R-4, the answer is R-56! Performed correctly, on an incremental basis, the answer is a more sensible R-16, since the incremental change from R-15 to R-16 costs 6¢ and yields exactly 0.6¢ savings per year.

$$\frac{1 \times 6000 \times 24}{15} \; \frac{\text{ft}^2 \times {}^\circ\text{F} \times \text{day} \times \text{hr/day}}{\dfrac{{}^\circ\text{F} \times \text{hr} \times \text{ft}^2}{\text{Btu}}} = 9600 \text{ Btu annual loss}$$

$$\frac{1 \times 6000 \times 24}{16} = \frac{9000 \text{ Btu annual loss}}{\;} \quad 600 \text{ Btu annual saving}$$

$$\frac{600 \text{ Btu/yr}}{1{,}000{,}000 \text{ Btu/MBtu}} \times \$10/\text{MBtu} = 0.6¢ \text{ annual cost saving}$$

Incremental cost-optimizing requires that energy-conservation measures, added passive glazing, south-window insulation, enhanced passive storage, and active solar systems all be arranged in descending order of cost effectiveness. Each is applied until the next measure decreases overall building cost effectiveness.

3. Passive elements in direct-gain systems can cause a lot of difficulty, because they are not interrelated as closely as are the elements in an active system. Suppose the client for a new house wants a lot of glass on the south, not only for passive gain, but for architectural reasons. The first step is to install thermal protection for the windows so that they do not become a net loss thermally. Having done this, the designer now must check to see if added storage is worthwhile. The only way to do this is to simulate the yearly performance of the design with and without storage, and compare the performance improvement to the cost of the storage. It is not unlikely that the optimum would be *no* added storage.

This is a sure sign that there is too much glass in the building; but the client wants the glass, and you, as a responsible designer, have provided ways to dump excess heat and control losses. The moral is that passive systems are not indivisible, and that their elements must be considered separately in practice.

4. If a Trombe-wall system is designed and cost-optimized as a predetermined unit, the designer is apt not to ask the crucial question: Is the storage component worthwhile? This is another example of the fallacy in not using incremental cost effectiveness. It is quite possible that removing the Trombe-wall, and substituting simpler heat-dump measures, along with night insulation, would improve the overall cost effectiveness of the design. It is likely that a reduction in glazing would also prove worthwhile. The liberated money could then be spent upon better energy conservation, thereby reducing the building's auxiliary demand for less money. While this is a hypothetical example, such analysis is essential. Remember, if you don't ask the right questions, you won't find the optimum systems.

5. In optimizing unitized systems such as active systems, it is tempting to set all the variables constant and vary one until a sub-optimum is reached, then repeat the process in turn for each variable. This is a very time consuming task and usually gives the wrong answer. A classic example is the load heat exchanger and store optimization problem. The next section (Guidelines for Designers: An Example of Optimizing) shows the results of a simplified one-day analysis of a system. If the load heat exchanger is arbitrarily sized too small, the optimum store size is reduced. Although not directly dealt with in the problem, we know that undersized storage raises collector operating temperature and reduces the energy harvest. Since the load heat exchanger cost is usually much lower that that of storage, it pays to oversize the heat exchanger, optimize storage, and then back down to a smaller heat exchanger if necessary. In general, oversize inexpensive elements, optimize the expensive ones, and then resize the inexpensive ones.

6. Watch out for non-linearities in component costs. Storage tanks, for instance, come in modular sizes, resulting in a "stepped" cost-to-size function. Obviously, the largest tank in each modular size is the most cost effective, if the cost of space or of burial is not an important factor. The same tank design might cost $1.00 per gallon in large sizes and $2.00 per gallon in smaller ones. Likewise, shell and tube heat exchangers tend to jump in price when the diameter is changed.

7. Pumps and fans introduce a new element in optimizing—parasitic losses. The value of these losses must be related to the value of the energy displaced by the system. For instance, if the displaced energy is

gas, parasitic losses should be multiplied by the ratio between electricity and gas costs. Cost, incidentally, has become a more reasonable measure of value since the deregulation of fossil-fuel prices. Remember that a large percentage, perhaps 40% to 60%, of the parasitic energy is delivered to the fluid as a result of increasing its kinetic energy (which is the purpose of pumping in the first place). This energy is in turn given up to the walls of the piping or ducts, replacing skin losses. Also, if the motor is in a heated space, the rest of the energy is useful during the heating season.

As long as electricity is treated as a commodity worth no more than its cost, straightforward analysis can yield optimum pump and fan sizes. The C.O.P. (coefficient of performance, i.e. solar energy provided divided by electric energy used) is more or less irrelevant in solar system design, as long as the electricity is appropriately valued. The following is a simple example of a pumping energy tradeoff:

Question: A system in the northeast, consisting of 3000 ft^2 of collector is expected to deliver 150,000 Btu/ft^2 of heat annually, assuming a flow rate of 0.5 gpm of water/20 ft^2 of collector. At that rate the F_R is 0.91; the manufacturer states that the F_R would be increased to 0.94 if the flow rate is doubled. What is the extra pumping power which could be added to achieve this higher flow rate, assuming half the power is usefully added to the fluid, and that electricity costs twice as much as the fuel displaced by the system?

Answer: The annual harvest would increase by $(0.94 - 0.91) \times 3000 \times 150{,}000$, or 13.5 MBtu if the flow rate is doubled. Since the cost of electricity is double but only half is wasted, electricity is effectively the same cost as the fuel displaced. 13.5 MBtu = 3955 kWh. If the pump runs an average of 4 hours per day for 365 days, or 1460 hours, we could just break even by adding 2.7 kW of pumping power.

It should be added that a complete analysis would include the capital cost of larger piping, since such an investment might save enough pumping power to pay for the added piping.

8. Sizing collectors on the basis of numerous optimizing curves found in the literature can be highly misleading. Such curves usually make many assumptions about the system, deviation from any of which may change the optimum substantially. Optimizing curves are often very flat, which means that they do not offer much useful information. In the author's experience, residential systems tend to have a constant cost effectiveness over a wide range of system sizes. Cost effectiveness goes down with the smallest systems, such as retrofit water heaters; or the largest (relative to the load), such as space heating systems delivering over 80% of the net load. Remember, errors in projecting the load can

make the fine-tuning of collector size a fatuous exercise. In a case such as that cited in Chapter 8, where the water heating load was overestimated by a factor of 6, the designer can be mightily embarrassed when a 50% solar system, plus or minus 20%, becomes a 300% solar system, plus or minus 100%.

9. As a general rule, backup systems should be optimized before the solar is considered. The whole purpose of adding solar is to minimize the use of backup energy. If this can be done by minor adjustments and improvements to the backup system, with quicker paybacks than that of the solar, obviously such measures take first priority. If the backup, like the solar, functions as a "base-load" energy plant, it must be jointly optimized with the solar (this would be the case with a heat pump, unless the heat pump preexisted the solar to provide cooling).

10. As one pays more attention to optimizing, it becomes more puzzling, because other issues enter unrelated to simple economics. The owner may want to add a microprocessor just for the fun of it, putting many control modes at your disposal (providing the sensors deliver good information to the processor). The owner may choose electricity over oil because it is "cleaner" (for him, not necessarily for the society at large). Suddenly many problems in relating backup and solar are solved for the designer; and the solar system becomes much more worthwhile. The owner, by spending more on backup energy, can now justify economically the capital cost of an expensive solar system—a most curious phenomenon!

11. At some point (often around 70% to 80% solar) the addition of enough solar input justifies the replacement of an expensive furnace, boiler, or heat pump by simple electric resistance heat, liberating capital for the further expansion of the solar system. Utilities have complained that higher than normal rates are appropriate for such installations, at best a debatable point. If higher rates *are* imposed, this will definitely influence the solar fraction at which the changeover to electric resistance heat takes place, if at all.

Suppose you have decided, because of the large solar fraction, to abandon fossil fuels and switch to resistance heat. The displaced energy is now much more expensive. If the system payback is refigured with electricity as the displaced energy, it will look dramatically better. Before congratulating yourself too enthusiastically, remember that the principle of incremental cost effectiveness also applies to the choice of backup.

12. By this time an experienced designer may be nervously asking: "All this optimizing on an ordinary consulting fee?" Of course not. Many of these steps can be performed carefully once for each typical system

and held in the form of rules of thumb. New situations should be examined carefully, however. What is a new situation? The list below describes some changes which might induce a reoptimizing of certain elements.

- A large change in the relationship between $F_R\tau\alpha$ and F_RU_L will lead to different optimum store sizes.
- Hydronic backup, which forces the system to run at elevated temperatures, might cause a change in the storage-collector ratio.
- Storage insulation is highly contingent upon environment temperature, and whether the lost heat is useful. If so, the optimum insulation is for summer conditions, to prevent overheating the space from storage.
- The nature of the yearly load curve has an effect on the optima. A continuous load such as hot water tends toward a smaller storage-collector ratio.
- The daily load curve also affects optima. If there is a consistent, large draw during collection, the store can be a bit smaller (this would be typical in a water heating installation). This is equivalent to saying that a system which is highly load-following requires less storage.

GUIDELINES FOR DESIGNERS: AN EXAMPLE OF OPTIMIZING

The choice of a load heat exchanger is not always under the control of the designer. For instance, in a warm-air heating system, where solar energy is delivered to the airstream by a duct coil, there are very few good coils available which combine all-copper construction, very low friction losses and large surface area. In addition, the size of the coil merely controls the *effectiveness*; the limit on the rate of discharge is set by the flow rate of the air. If only a small airstream is being delivered, even a 100% effective duct coil might not be able to deliver energy at a desirably high rate.

In order to make clear the interaction of store size and heat exchanger discharge rate in an example, it is helpful to make a few simplifying assumptions.

1. Although the purpose of enlarging the store and heat exchanger is to increase collection, it is possible to perform the optimization without taking into account the increase. One can set the net gain from the system at a reasonable amount typical of the month during which the system is to be optimized—in this case, January; 1000 Btu per ft^2 collector per day is assumed.

2. It is assumed that the load heat exchanger is working continuously —that is, that there is a continuous load. In fact, even if there isn't, the results are similar, providing that the system is not approaching a 100% solar contribution.

3. Both room temperature and ambient temperature are kept constant.

4. The load heat exchanger runs only after collection, for a fixed number of hours. This is a reasonable assumption for many residences, where the daytime load is satisfied by solar gain and occupancy heat. For the example, an 18-hour discharge period is assumed.

In words, the dynamics are as follows: the (fixed in the example) charge of solar energy enters the system and raises the temperature of storage by an amount equal to the charge divided by the capacitance of the store. Discharge is determined by a simple exponential function which tells us what percent of the peak energy is discharged during the fixed discharge time period (18 hours). This function depends upon the system *time constant*, which is the storage capacitance divided by the *heat transfer rate* of the load heat exchanger. In the example a 400 ft^2 collector is used in a house having an air circulation rate of 1000 CFM, with a load heat exchanger (duct coil) having an effectiveness of 74%. The heat transfer rate of the coil will be 1000 × 1.08 (the capacitance of 1 CFM of air flowing for one hour) × 0.74 = 800 Btu/°F/hour. With a 400 ft^2 collector, this equals 2 Btu/°F × hr × ft^2 collector.

As can be seen in Fig. 9.2, reducing the system time constant will result in a more thorough discharge of storage, thereby lowering the temperature of the system at the beginning of collection, and enhancing collection. But there are two ways to accomplish this: if the heat exchanger output is fixed, one must reduce storage capacitance; while if the store is fixed, one must enlarge the heat exchanger. The effects of the two strategies are opposite; the first causes the average operating temperature of the collector to *increase*, since the reduced store raises the average collector temperature more than the reduced initial temperature lowers it. The example shows that doubling the output of the load heat exchanger (which is likely to be done by increasing the air flow rate and the fan energy rather than enlarging the coil itself) results in a larger store and a much lower collector operating temperature.

To find the optimum, one must compare the incremental value of the reduced collector temperature with the cost of the added storage. This really requires an annual simulation, but for this simple example, assume that the heating season consisted of 100 days identical to that in the example. If the collector operated for 7 hours each day, a 1°F drop in its average temperature would be worth 700 × the collector $F_R U_L$,

ST		TC	DTS	% Discharge	TS_{max}	TS_{av}
Gallons	Btu/°F	—	°F	—	°F	°F
1	8.33	4.2	120	0.99	121	61
2	16.66	8.3	60	0.88	68	38
3	25.0	12.5	40	0.76	53	33
4	33.33	16.7	30	0.66	45	30
5	41.66	20.8	24	0.58	41	29
6	50.0	25.0	20	0.51	39	29

Constants: Net gain = 1000 Btu UHX = 2 $\frac{Btu/}{hr \times {}^{o}F}$ T_{ra} = Constant

H = 18 hours

ST		TC	DTS	% Discharge	TS_{max}	TS_{av}
Gallons	Btu/°F	—	°F	—	°F	°F
1	8.33	2.1	120	0.99	121	61
2	16.66	4.2	60	0.99	61	31
3	25.0	6.3	40	0.94	42	22
4	33.33	8.3	30	0.88	34	19
5	41.66	10.4	24	0.82	29	17
6	50.0	12.5	20	0.76	26	16

Constants: Net gain = 1000 Btu UHX = 4 $\frac{Btu/}{hr \times {}^{o}F}$ T_{ra} = Constant

H = 18 hours

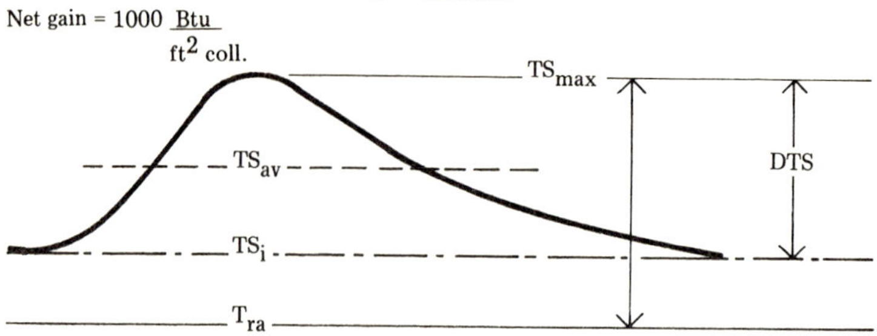

Note: All quantities normalized to one ft^2 collector

DTS = rise in store temperature

TS_{max} = maximum store temperature (measured above T_{ra})

T_{ra} = return air temperature (assumed constant)

$\%\ \text{discharge} = \frac{DTS}{TS_{max}}$

H = hours of discharge (assumed 18 hours)

$TC = \text{system time constant} = \frac{ST}{UHX}$

$ST = \text{storage capacitance } \frac{Btu}{{}^{o}F}$

$UHX = \text{load heat exchanger heat transfer rate } \frac{Btu}{\text{hour} \times {}^{o}F}$

$TS_{av} = TS_{max} - \frac{DTS}{2}$

Fig. 9.2 Analysis of system output for varying store sizes and for two different heat exchanger sizes. The larger heat exchanger results in a larger optimum store size. This trend will continue until the added output resulting from the larger store is not worth the cost of the larger heat exchanger. Unfortunately, many designers draw the opposite conclusion: since larger heat exchangers result in larger stores, reduce the size of the heat exchanger and save on the first costs!

which might be 1.00. At $10.00 per MBtu, the 1°F would then be worth 10¢/ft^2 collector, using a 15-year simple payback ($700 \times 1.0 \times 1000 \times 15/1{,}000{,}000$). If the incremental cost of adding storage were 40¢/gallon, we could add storage up to the point where a gallon reduced the average temperature by 4°F. In the first run, with the smaller heat exchanger, the optimum would be about 2.5 gallons per ft^2 of collector, while in the second it would be closer to 3.0 gallons per ft^2 of collector. But notice that enlarging the heat exchanger dropped the average from about 35°F to about 23°F.

This crude example should be supported by a detailed annual simulation. The program used must allow independent adjustment of the load heat exchanger and the store. It should be noted that, except in the coldest climates, the discharge rate from storage is often limited by the heat loss rate of the building—you can't dump more heat into the building than it loses without raising its temperature. This puts an upper limit on the useful output of the heat exchanger. Likewise, a low limit on storage discharge can also put an effective upper limit on heat exchanger size.

10

DESIGNING A SIMPLE ACTIVE SOLAR HEATING SYSTEM

THE DESIGN OF DRAIN-BACK SYSTEMS

Liquid systems, with all their advantages of compactness, flexibility, efficient heat transfer, and freedom from leakage, have their own disadvantages: serious damage if leakage *does* occur, and freeze-up problems. The former can be solved with proper workmanship, just as any plumbing system can be made watertight. The latter requires elaborate care and is the cause of most of the overdesign of systems.

Antifreeze can be an abominable, expensive, nuisance, but in order to use tap water as a working fluid in freezing climates, the water must be drained adequately from the collectors before freezing can occur, or the collectors must be able to withstand freezing. While the latter idea is being pursued, using rubber tubing either outside or inside the water stream, most collectors use metal piping and must be drained. (Even if freezing is safe, it requires a lot of energy to thaw the resulting ice.)

To understand the principle of a drain-back system, picture a collector mounted high on a roof, connected at both top and bottom through continuous piping to a sealed tank in the basement, with both pipes entering the tank below water level, as in Fig. 10.1. Assume for a moment that the water is cold and the system is at sea level. In a pressurized system (typical of any household hydronic heating system), the top of the system is above atmospheric pressure. If a vent is opened at the top, water will squirt out.

If, however, a vent is opened in the pressurized air above the top of the tank, the system pressure is relieved and atmospheric pressure in the system drops to the *level of the water in the tank*. But what happens to the water in the system? Will it cascade down as a result of relieving the system pressure? If the column of water is higher than 34 feet, at which

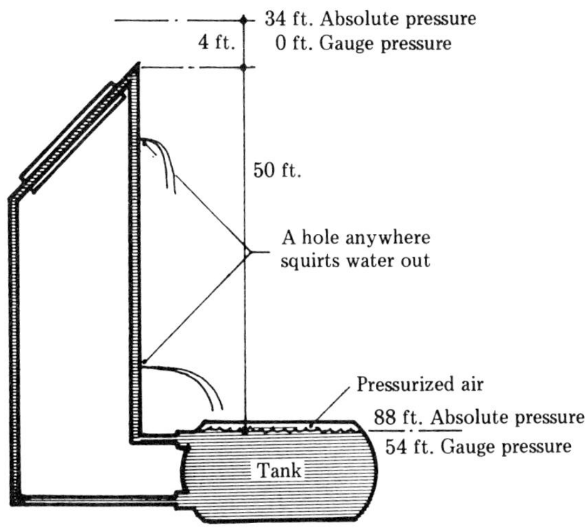

Fig. 10.1 A fully pressurized collector-store loop containing water.

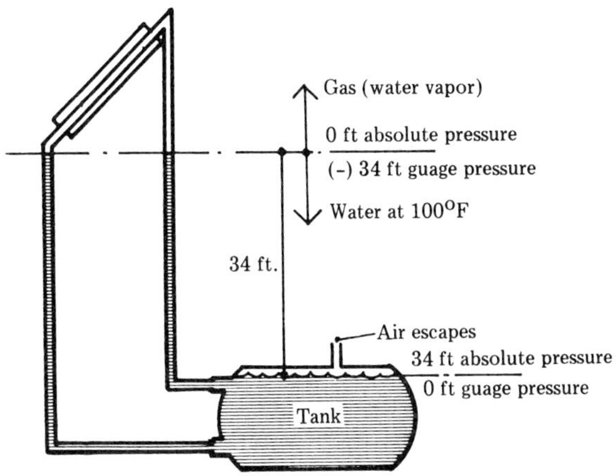

Fig. 10.2 When pressure is relieved from the tank, with the loop outlets under water, the loop becomes a water barometer. If the loop is high enough, the water at the top will vaporize, at a height dependent upon the altitude of the site and the temperature of the water.

height its weight balances that of a column of atmosphere, the upper part will vaporize (Fig. 10.2); but the lower 34 feet will stand in the pipe, simply because there is no way air can get in—remember that both pipes terminate under water. We have invented the water barometer, a refreshing change from the wheel. If a vent is now opened anywhere in the pipe, even at the bottom, air will rush into the hole and the water in the pipe will empty back into the tank (Fig. 10.3). The vent does not have

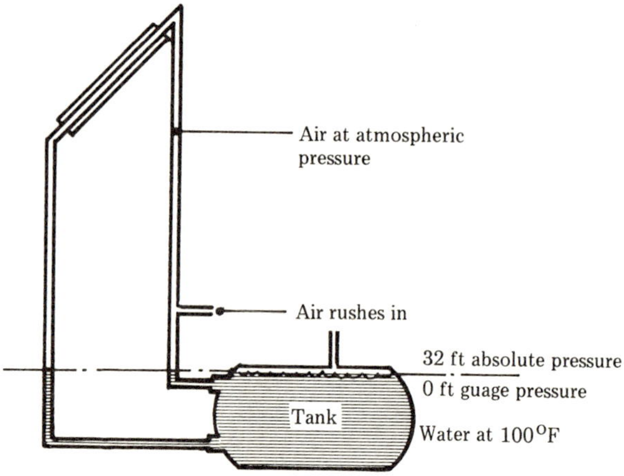

Fig. 10.3 Because the loop is below atmospheric pressure, opening a vent anywhere in the loop causes air to rush in.

to be at the top, since the entire column is at a pressure below that of the atmosphere outside the pipe, and air will try to flow in at any point. Another way of accomplishing the same effect is to lift one end of the pipe out of the water. If our initial column had been lower than 34 feet, the pipe would initially have been full of water, with no air at the top. Upon allowing air to enter the pipe near the bottom, most of the water above the vent will siphon over the top and down the other side.

To generalize this simple model of a siphon, recall that the pressure at which water vaporizes depends nonlinearly upon temperature. And, the other way around, the temperature at which water vaporizes depends upon the pressure, as any mountain-dweller knows. Fig. 10.4 shows the relationship between the vaporizing pressure and the temperature at sea level.

In designing a system which uses a siphon, or in figuring the pressure on the intake side of a pump to prevent cavitation, you must design for the worst condition, that is the hottest water going through the pump; and for the altitude of the system. The chart shows the pressure not in psi, but in feet of head. This is a great convenience, since pumps are so calibrated, and since the vertical dimensions of the system are of great importance, and are also measured in feet. The number of feet of water which can be supported at a given altitude and a given water temperature will be referred to as the "atmospheric head."

Suppose now we examine the simple solar system diagrammed earlier. It closely resembles the siphon diagrams above, except that collectors are mounted at the top of the system, and a pump is at the

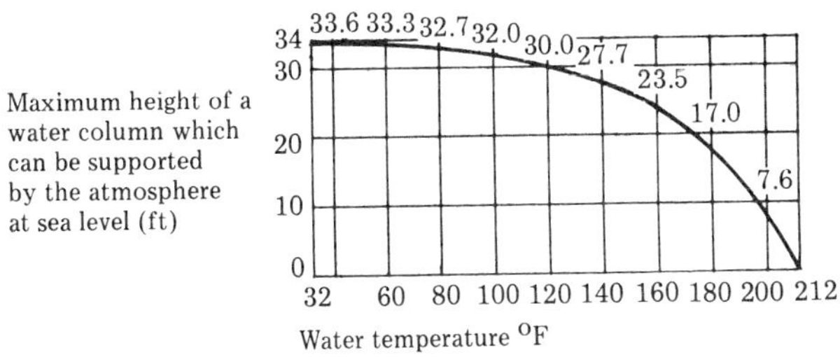

Fig. 10.4 The pressure at which water vaporizes at sea level is also the height of the column of water which can be supported by the atmosphere. As the operating temperature of the water in the collector loop rises, the tendency for it to vaporize increases, an important factor in dealing with pump NSPH.

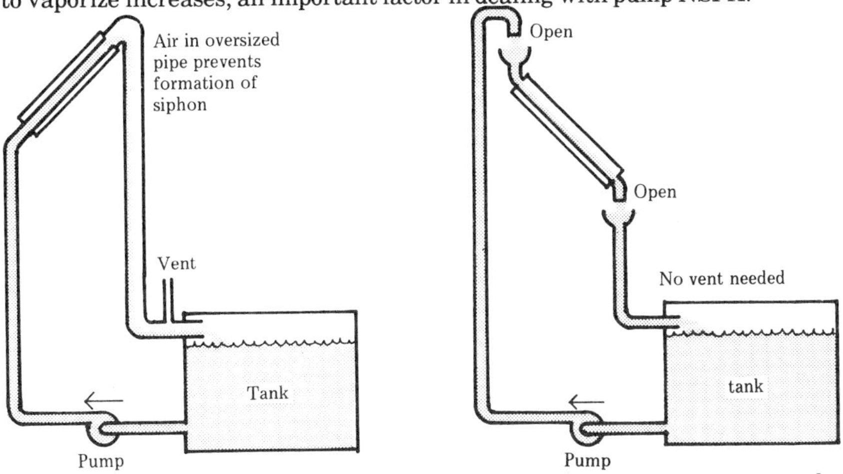

Fig. 10.5 An oversized return pipe, or a trickle-down design, prevents the formation of a siphon. After fill-up, the pump lifts water the full height of the system at all times.

bottom. A permanent vent is located at the bottom as well. At rest, the water in the system fills the piping to the level of the water in the tank, flooding the pump and guaranteeing that it will have water in it when it starts. There is no water in the collectors, which means that they will not freeze on cold nights.

When the sun comes out, warming the collectors, the controller starts the pump, forcing water up the supply pipe until the collectors are full. The pump, to accomplish this task, must lift water the full distance from the water level of the tank to the very top of the system, a dimension known as the system's "gravity head." If the pump uses almost all of its available power in doing this, only a trickle of water will flow over the top, and the system will be stable. Even with a substantial flow, if the

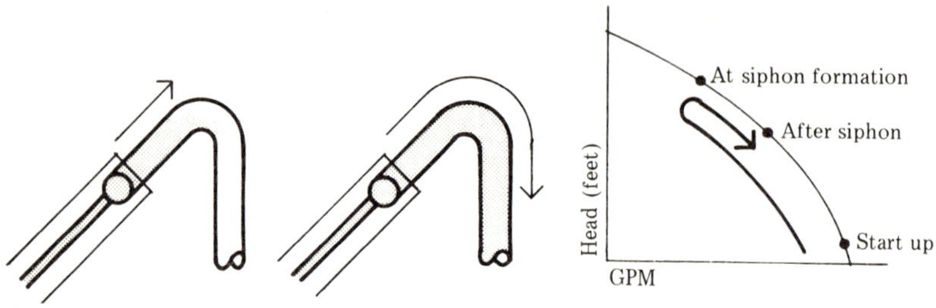

Maximum pressure on pump, minimum flow, just before siphon forms

Pressure begins to be relieved, flow increases, after formation of siphon

Schematic pump curve, showing how pump operating point changes at start-up

Fig. 10.6 Siphon formation in a properly designed drain-back system.

return pipe is large, or if the water runs down an open surface (as in a trickle-down system), the pump will work constantly to lift water the full gravity head of the system (Fig. 10.5).

But suppose the pump pushes a noticeable flow over the top into a normally-sized pipe. Quickly, the pipe becomes full of water, driving the air before it, down and out of the pipe, thereby forming a siphon. This happens if the static pressure in the pipe exceeds about one foot per hundred feet (about 0.4 psi), according to one source; but it surely will happen at 2 or 3 feet (about 1 psi) according to the author's experience, even if several banks of collectors are present, with numerous high points.

With the siphon formed, a remarkable thing happens; the flow rate begins to go up, because some of the gravity head on the pump is relieved by the "new" head made available by the siphon. Once over the top, the water can gain energy by falling. Because water acts in some ways like a solid, this energy can be transmitted back over the siphon to the water climbing the supply pipe, and used to help that water over the top. Relieved of this load, the pump can change its operating balance between potential and kinetic energy output, substituting gpm for feet of head (Fig. 10.6).

If the return pipe is shorter than the atmospheric head (neglecting for the present the effect of pipe friction), water will cascade all the way to the bottom of the system without vaporizing, and the pump will "see" almost no gravity head (except the small one between the return pipe and the water level in the tank. It will behave then exactly as if the system were a closed loop. If, however, the return pipe exceeds the atmospheric head, the water will vaporize somewhere in the return pipe, breaking the siphon (Fig. 10.7).

Suppose we redraw the system with the return pipe cut off exactly at the point where the water vaporizes. The remaining return line, and

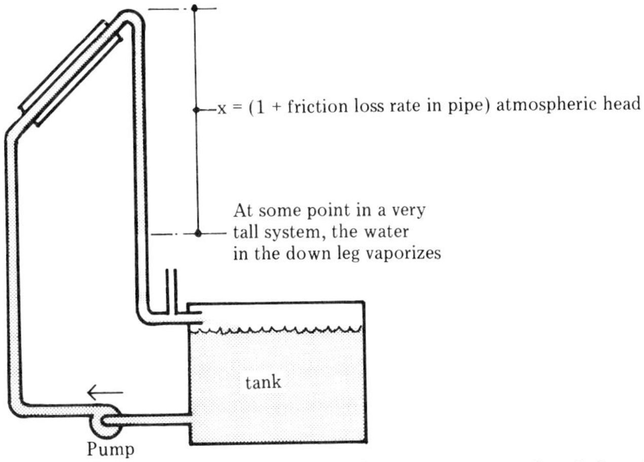

Fig. 10.7 In a tall system, operating with warm water, the siphon is likely to break somewhere in the downleg. This should be taken account of in selecting the pump and designing piping, but otherwise has no effect upon operation.

the corresponding length of supply line form the siphon. The pump's job, then, is to lift water to the level of the end of the return, and to deliver it at a rate sufficient to replace the water lost out the return pipe.

In order to accomplish this, the pump must overcome a *gravity head* equal to the distance between the water in the tank and the end of the siphon; and it must overcome the *static head* generated by the friction in the piping all the way from the tank to the end of the siphon.

As the water falls down the return side of the siphon, it produces friction, which in turn decreases the energy of the moving water. We said that the water vaporized at a point in the return when it had fallen by a distance equal to the atmospheric head, neglecting friction losses. Adding back the friction losses (and assuming that they are constant in the vicinity of the vaporizing point), the downleg of the siphon will be:

$$(1 + \text{rate of friction loss per foot}) \times \text{atmospheric head} = \text{length of siphon.}$$

This relationship will help us to establish the "system curve" when we size the pump.

What happens when the pump stops? In the "short," low-head system, air enters at the bottom vent, and the system drains like any siphon. Water will flow backward at a high rate through the pump. In the taller system, the sudden reduction in pressure can cause the water at the *top* of the system to vaporize. Upon vaporizing, the water no longer forms a siphon; and, if flexible hoses are used, they may collapse under atmospheric pressure. Both effects cause the system to drain very slowly, since air must now bubble up from the bottom on both sides before drainage can occur. It is exactly like emptying out a narrow-necked bottle by holding it upside down: the system "chug-a-lugs."

To prevent this, a simple valve can be installed on the supply side, near the top, but indoors. (It must be closer to the top than the minimum atmospheric head of the system, which occurs when the water is hottest.) The device is a "vacuum relief valve," the kind which is used on domestic water heaters. It opens immediately whenever the pressure in the water falls below atmospheric pressure, letting in air. The siphon is saved, and water runs rapidly down both sides of the system.

More experimentation and measurement are needed to determine exactly under what conditions drainage is slowed enough to warrant the installation of such a valve. In the author's experience, systems having only one bank of collectors and a gravity head less than 35 or 40 feet, do not need a vacuum relief valve. In any case, *do not* put one on the return side! This destroys the siphon all the time, and thoroughly messes up system operation. For systems without vacuum relief valves, it is a good idea (borrowed with thanks from Acorn Systems) to insert a length of tubing on the inside of any flexible hoses which are longer than one inch, to prevent their flexing during drain-back.

Because of the continual presence of oxygen in the open system, it is appealing to search for a system which is closed, but which nevertheless drains back by exchanging a trapped volume of oxygen-free air between a holding tank and the collectors. (The oxygen leaves the air in a closed system rapidly, by depositing scale on the surfaces of the piping). Such systems are substantially more complicated than the open drain-back system described above. Despite the advantages in allowing the use of standard cast-iron parts and avoiding evaporation problems, the drainage cycle is less positive, and the design requires much more care than that of an open system. In particular, attention must be paid to the change in the volume of the trapped gas as its temperature rises and falls.

Two other drainage-type systems, both called "drain-down" systems, will be touched on briefly. A popular design runs domestic water to be heated directly into the collector, where it is circulated either by gravity or by pump into storage. As water is drawn, new water flows in and runs through the collector. In a similar design, space heating is done, with water under house pressure used in the collector either as a heat transfer fluid replacing antifreeze, or directly from a pressurized storage tank.

In both cases, the onset of freezing conditions in the collector causes a differential controller to energize or deenergize electrical dump valves, which simultaneously hold back line pressure and let the water out of the collectors into a drain.

The design, while very simple and widely used, has several faults. First, in the domestic water-heating system, a constant supply of new domestic water is run through the collectors. If the water is soft and has a neutral pH, this is fine. But ordinary city water can be a bad brew to run through the thin-walled tubing characteristic of many collectors. The worst fault is that drainage is dependent not, as in the drain-back system described earlier, upon the pump's stopping; but upon the combined action of an electrical dump-valve and a differential controller. If either fail, the system is very likely to freeze up. The numerous failures of such systems have given a bad name to all forms of drainage systems. However, continued development will undoubtedly improve the reliability of the key controls, and as long as they are properly maintained, such systems will continue to work and be popular. The author would not consider such a system in a cold climate or on a large project.

(The attempt to distinguish between the two types of drainage systems by using the terms "drain-down" and "drain-back" is a classic example of 20th century America's semantic desperation. The author, a typical, semantically desperate American, has observed the distinction.)

Drain-back systems leave the collector subject to total stagnation, but in an empty state. As with a pressurized loop, it is possible to introduce some kind of heat-dump measure, such as an in-line radiator, and then cross your fingers that the heat dump doesn't break down on a hot day. Best is to choose a collector which can survive stagnation on its own, without such a support device. If the system starts up during stagnation, steam will form, but it will merely escape from the vent without any problem. Very low-loss collectors can be damaged by thermal shock if started up when hot. This is another good reason for using relatively high-loss collectors, perhaps with a fiberglass cover and a painted absorber rather than a glass cover and selectively coated absorber.

One big advantage of the drain-back design occurs in systems with lots of outdoor piping. In antifreeze or drain-down systems, the fluid in the outdoor piping can reach ambient temperature. At start-up, this fluid can enter the collector, cooling it off and causing start-up cycling; but it definitely will enter the building, causing a thermal shock to piping, perhaps setting off low-temperature alarms, and subtracting heat from the system. In a drain-back system, only the pipe itself can get cold. Copper has a specific heat of .09 and the tubing weighs less than the water in it. The water filling the cold pipe loses relatively little of its heat to the pipe, thus eliminating these problems.

Under what circumstances can a drain-back system freeze? First, if the piping does not pitch—there have been many examples of this, again giving drainage systems a bad name. Second, any system will freeze if the collector pump runs continuously and unnoticed (a simple roof-mounted light could attract attention to nighttime operating systems). A reliable differential controller is important, possibly with a back-up indicator device and/or an automatic shut-off at 36°F. Finally, frost could form in the vent pipe, if it extends beyond a heated space. The solution is to see that the vent pipe does not extend beyond the heated space.

Control of a drain-back collector is the same as that in an antifreeze system. A differential controller starts the pump whenever a carefully located sensor attached to the absorber plate of the collector (see Chapter 4) registers a temperature some amount higher than the temperature at the middle or bottom of the storage tank. It turns off at some lower differential—15°F and 5°F are recommended turn-on and turn-off differentials. It is crucial that the collector sensor be mounted to a separate section of absorber, unless the flow rate is so low or the fin efficiency so low that the sensor can be mounted on the absorber itself. A high temperature limit is usually needed, sensing tank temperature. There is no particular reason for limiting the maximum operating temperature at the collector, unless a low-loss collector is used, in which case the system should be prevented from filling if the collector exceeds a reasonable temperature—perhaps 250°F or so. It is important in larger systems to include an adjustable time delay to prevent the pump from starting during drain-back. A single-phase motor can start up in reverse, and a large motor can be damaged if started while running backward.

DRAIN-BACK SYSTEM DETAILS

General Rules:

One of the advantages of drain-back systems is their simplicity, and it is a design goal used by the author to minimize the use of valves (Fig. 10.8). This runs contrary to usual good plumbing practice, and it is often difficult or impossible to convince a good plumber *not* to install valves where they are not absolutely needed. The main reason you don't need them is that the system is not under pressure. Suppose, for instance, that the pump must be removed. The one gate valve required, located between the pump and the tank, will prevent the tank from draining when the pump fittings are opened up. The other side will spill a little water onto the floor and then will be dry. It is incidentally very important not to leave the gate valve between the pump and the tank in an adjustable position. If someone inadvertently closes the valve, the pump will cavitate and in some cases burn out. Solder it open, or remove the handle.

Another general rule in detailing is to be sure the entire length of

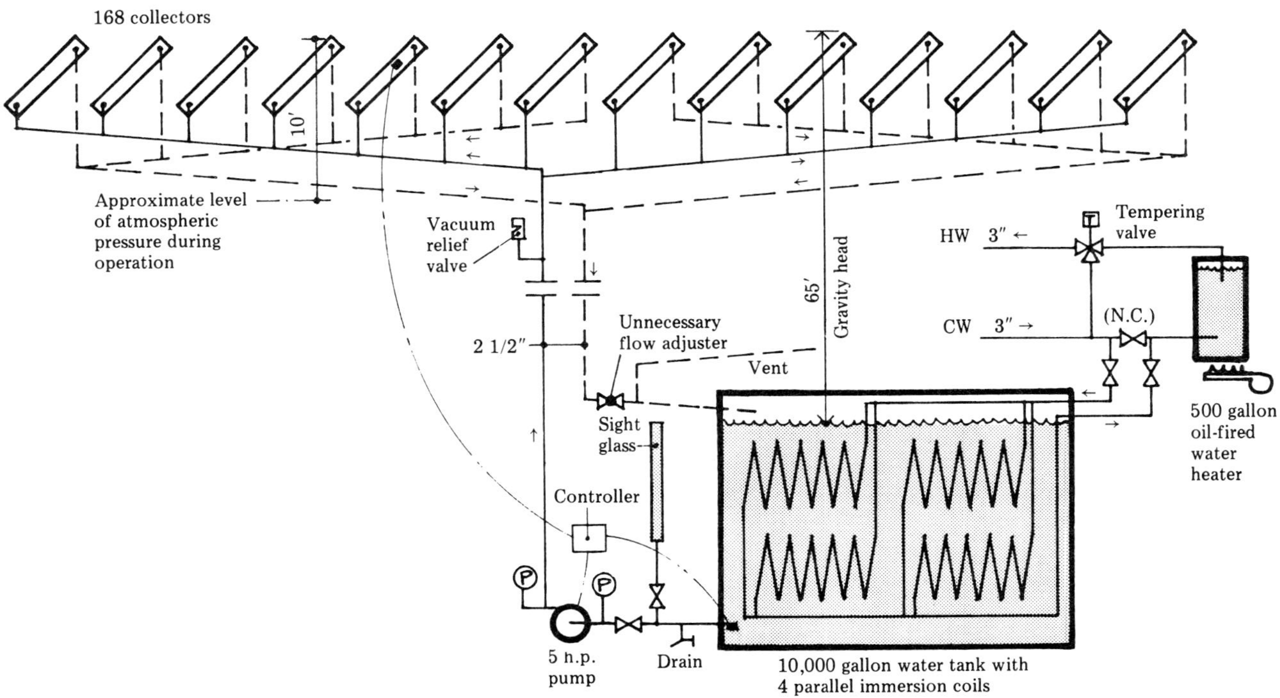

Fig. 10.8 The author's design for the system at Oak Park housing in Lewiston, Maine, built under a HUD demonstration grant. Experimentation showed that the vacuum relief valve should never be placed on the return line (the valve is unnecessary in systems with smaller gravity heads). Every valve and accessory in the DHW and collector loops is shown in the diagram. A secondary antifreeze ventilation air preheating loop, operating through a heat exchanger, is omitted to show how few devices are necessary in a successful drain-back DHW system. It is disconcerting to observers that opening a joint anywhere in the collector array during operation results in the intake of air (the entire upper part of the system is below atmospheric pressure).

both the supply and return piping pitches back to the tank. The idea is hard to get over to heating installers, because with pressurized systems, you merely have to install drain valves and vents wherever you have to trap the line. In drain-back systems, the entire line must pitch, just like a drain pipe. Even a short length of what seems to be horizontal pipe will cause drain-back problems. And remember that the building may in the future settle, a further encouragement to pitching the lines well.

As in any solar system, all piping must be insulated. This runs against the grain for most plumbers, although heating installers are used to some insulation. It is not uncommon to see a solar installation with insulated pipes, some of which never gets above 100°F, next to uninsulated water piping which is always at 140°F. The plumber did not include insulation in his price for the building, but did in his price for the solar system.

Because the system contains dissolved oxygen at all times, there must be no, repeat, no steel, cast iron, or aluminum in contact with the water stream. There may be occasional exceptions to this: one can, as in a cooling-tower system, install a water treatment system. But, chemical treatment is only as good as the person maintaining it. Homes are sold on the average every 7 years, and sophisticated maintenance procedures are likely to get lost in the move. Even in larger buildings, it is not uncommon to find water-treatment systems removed or badly maintained. Reiterate: no steel, cast iron, or aluminum.

Other general issues will be covered later, since they apply to all systems. There are several specific design problems which arise with drain-back systems, and which need detailed discussion.

Pumps

Pumps must be all-bronze, stainless steel, or plastic, and have no aluminum, iron, or steel in the seals or other places which touch the water stream. Its operating curve must satisfy the demanding requirements outlined in the pipe and pump sizing procedure found later in this section. Often, a single pump cannot be found with the proper characteristics, especially in residential systems where a small flow rate must be coupled with a high head. Here, two or more identical pumps may be used in tandem (one behind the other). The pump heads add, but not the flow rates, and you mimic a single pump with the desired high-head, low-flow characteristics. In addition, some new pumps are coming on the market with these characteristics.

Frequently, the pump with the proper curve will operate at a high RPM. This raises the question of cavitation and NPSH (Net Positive Suction Head). Since the water level of the storage tank is always at atmospheric pressure, there is intrinsically a small head on the suction side of the pump. This would cause no problem for a "positive displacement pump," but such pumps cannot be used, since water must flow backward through the pump during drainage. A well pump might work; the author has never tried it. Usually, ordinary centrifugal pumps are the best choice. If the pressure on the inlet side of the pump is insufficient, the pump will cavitate—that is, it will vaporize the water by the action of the impeller. This not only pits the impeller, it will in a high-speed pump stop the flow of water, causing some designs to overheat and burn out.

Each pump has a requirement for a specific NPSH. This requirement typically rises as the water temperature rises. The pressure on the inlet side is defined by the NPSA (Net Positive Suction head Available). This is basically atmospheric head plus the gravity head minus the friction loss, and can be calculated for any temperature and system configuration. Since the requirement goes up with temperature, the critical problem occurs with the hottest water allowed in the storage tank. Here is a design tradeoff. Letting the tank get hot will under some circumstances increase the carryover. But the hotter it gets, the worse is the cavitation problem. (Also, as will be seen, tanks with plastic liners cannot stand very high temperatures.) Cutoffs in the range of 160°F to 170°F usually will do the trick, although a range of 140°F to 150°F is often used in domestic water heating systems.

How do you find out what the pump NPSH is? The larger manufacturers of centrifugal pumps for heating use have for years published NPSH curves in their literature, but not for many of the smaller circulators. Makers of many of the sophisticated new plastic and stainless steel pumps either don't know or won't supply such information. Your solution is to make a test. Take the pump and fit it with hose connections. Fill a large bucket with water at the maximum store temperature. Place the pump below the bucket, run hoses from both the supply and return into the bucket, and turn the pump on. Now, raise the pump relative to the bucket (or vice versa) until the pump cavitates. Try to make the hose friction roughly equal to the friction on the intake side of the pump as actually installed. This procedure will not yield the exact NPSH, but will tell you what you want to know, namely how far below water level the pump must be mounted. Be conservative and add a foot, to account for variations in pump manufacture and possible constrictions in the intake line which you forgot.

In general, pumps with large inlets will not cavitate as readily as those with small ones. Condensate pumps are designed for this kind of application, but are hard to find in other than cast-iron construction.

Remember when reading pump data that the pump has only the gravity head between water level and the pump on the supply side when starting. The NPSH curves, as well as the pump curves themselves, often stop short of such low operating pressures. For this reason (the availability of data), as well as the requirement of preventing cavitation, it is well to mount the pump as low as possible.

One other note on pumps: some manufacturers do not publish accurate curves! While you are testing the pump, you might extend the hose on the supply end, get a helper to lift it, and find out the height above the pump at which water ceases to flow. This is the "shutoff" head, and should correspond to the head in feet shown on the end of the pump curve at zero flow rate. If it doesn't, write an angry note to the manufacturer.

System Venting

As described earlier in this chapter, venting at the top of the system is unnecessary until the gravity head exceeds about 40 feet. The bottom vent deserves attention and must be installed carefully. The best and safest approach is to install a short length of small diameter tubing at the top of the return, just before it enters the tank. If the return discharges without a fitting at the end and above water level there will be very little pressure remaining in the return at the vent, and water will not rise far into the vent pipe during operation. If you put your hand over the outlet and restrict its flow, pressure will be built up and water will rise in the vent, ultimately overflowing. Installing a fitting at the end of the return, or terminating the return underwater, will introduce such back pressure, requiring a longer length of vent piping.

Why not vent the tank, and let the return discharge above water level? As long as you are absolutely sure that the return will not accidentally discharge below water level, this detail will work. It is a risk that few experienced drain-back system designers take, however.

To prevent vapor loss through the vent, a small P-trap can be installed. The trap should be insulated thermally from the vent, and should be small enough in diameter to allow vapor to condense and trap the vent. Upon venting, the trap is sucked by the pressure—make sure this can happen! Another measure which works in certain situations is to extend the vent tube, pitching up, along a cool (but not freezing) surface, such as the outside below-grade wall of a basement. Vapor will cool and condense, and drain back to the tank. Neither of these measures will prevent the heat loss through vapor condensation; but both will prevent the necessity of making up tank water frequently, and will prevent damage to the surroundings from the condensation of the hot vapor.

Be sure the vent can never freeze, because the vapor will form a frost plug and prevent drainage. Thus, never run it outside.

Tank Design

Because the tank is always at atmospheric pressure, it can be built like a barrel. Essentially, all you need is a plastic liner and some means of restraining it when it is full of water. There are as many designs as there are brains to invent them.

- The Acorn tank consists of a carefully designed 3/8″-thick plywood cylinder, put together on the site (so each piece can be walked through a door). The shell is held together with steel bands, and the shell insulated. A plastic liner is placed in the tank, and a plastic sheet put over the top.
- The same basic design can be accomplished using a rectangular wooden tank, with framing holding up the plywood. The framing can be braced to the structure, reducing the cost. Likewise a steel oil storage tank with the ends cut off, or a piece of concrete sewer pipe can be used.
- Manufacturers are now offering prefabricated tanks, in some cases integrally insulated, with through-wall fittings. One of the problems with homemade tanks is that through-wall fittings leak, some earlier than others. It is so risky to install them that over-the-top piping is a must. With prefabricated designs, very convenient piping through the wall is possible.
- For larger tanks, it is often convenient to install concrete or masonry walls and insulate them on the inside, with a liner over the insulation. Remember that at least the inner layer of insulation should be able to tolerate the highest tank temperatures (use polyurethane); and that all the insulation must be able to withstand the water pressure (no fiberglass batts except on top). Be sure to provide weep holes through the concrete at the very bottom of the tank, so that if there is a leak, the water will not fill up behind the liner and float the insulation (it will also signal a leak). If the tank floor is below basement level, as it often is, provide a small pit at tank floor level, preferably with a sump pump, into which to empty the weep holes. The pump may also be located in this pit, if precautions are taken to insure that the pump does not get flooded during tank drainage.

To prevent vapor loss, the cover of the tank must be sealed to the liner. In addition, the cover must be impervious—any seams must be sealed, as well as holes made by fasteners holding the cover to the roof structure. If the return ends below water level, the tank itself must be vented exactly as is the return pipe. Piping running through the liner should penetrate above water level, and the holes should be sealed with silicone.

To make the sealing problem more difficult, the cover must be at least partly removable for inspection and adjustment of piping. Clips, closed-cell foam gaskets, or sealant which can be peeled off easily are all means to provide a removable seal between the cover and the liner. This is another reason why prefabricated tanks with built-in manholes are attractive, providing they can get through the available openings into the space housing the storage.

If the tank must be buried, it is very desirable to use a prefabricated design, with a manhole at the top. Bury it below the frost line, and extend the manhole with a section of culvert or other means of holding back the earth. Removable insulation should be placed over the tank at the bottom of the extended manhole which reaches the ground level. If such tanks are not available, revert to an antifreeze design and use a pressurized steel tank.

Some means must be provided to fill and drain the tank. Drainage is simple if a boiler drain is placed next to the pump. Filling can be accomplished using the same valve if a female-female fitting is added to the boiler drain to receive the male end of a hose. In fancy jobs, a permanent valved-off connection to a cold water supply can be included. Finally, water can be poured into the tank through the vent, especially if a female hose fitting is provided.

A convenient way to sense tank-water level is to provide a 1/2″ tee off the collector pump intake pipe, fitted with a shut-off valve. After the valve, clamp on a length of plastic or glass tubing which runs just up to the desired level of the tank. Open the valve to see water level. Unfortunately, if the filling point is connected to such a sight glass, water will be forced out the top during filling—a good reason for filling through the return line. By keeping the sight glass line open while filling (or opening it periodically), overfilling is impossible, since excess water will flow out the top of the glass. Don't put the sight glass on the pump discharge, for, if left open, it will allow the pump to drain the tank onto the basement floor. Keep the gauge valve closed to prevent vapor loss.

Other clever ways of measuring tank level can be devised. For instance, if the cover is allowed to float on the water, using a block of polyurethane as a support, an indicator attached to the center of the cover could project above the top insulation.

Pipe Sizing and Pump Selection for Drain-Back Systems

1. *Design Piping*
    a) Place pump outside tank, far enough below surface of water in tank to avoid cavitation.
    b) Use "over-the-top" siphon on pump suction to avoid fittings through tank waterproofing, unless tank has guaranteed built-in fittings.
    c) Use reverse-return piping in collector array to avoid "starving" some collectors.

- d) Make sure all piping drains back to the tank from a minimum number of high points.

2. *Calculate gravity head* as distance from top of water tank to high point of system
   - a) For systems with gravity head less than 35 or 40 feet, and with only one bank of collectors:
     - i) arrange return to discharge above water level;
     - ii) place vent on discharge pipe outside tank;
     - iii) devise a condensate collector for vent.

     Place *no other* vents, vacuum breakers, or relief valves in system.
   - b) For systems with greater heads, design as above, but:
     - i) add a vacuum breaker within an atmospheric head of top of the system, on the *supply side only*, located in a space which does not freeze.
3. *Establish a design flow rate*
   - a) Either make detailed performance trade-offs between higher collector F_R and extra initial cost and operating cost, or
   - b) Use rule of thumb of 1/20 gpm per square foot of collector.
   - c) Be sure to check collector to see if friction loss in collector manifold restricts design flow rate.
4. *Size Return Piping—Preliminary*
   - a) Use a roughly constant pressure drop for various pipe sizes.
   - b) Design so that total friction loss on the return side approximately equals the gravity head minus 5 feet.
   - c) Do not use pressure drops less than 1 psi per 100 LF.
5. *Resize Return Piping—Final*
   - a) If any of the return piping sized in step 4 has such a high flow rate that erosion becomes a problem, it must be resized upward until no erosion problems occur. Experts vary on their estimates of the flow rates at which erosion is a problem, but 4 to 6 feet per second is safe for small piping, and rates of 10 fps or so are acceptable with piping 2″ and above.
   - b) If the return piping must be resized, there is a good chance that the siphon will not reach the bottom of the system. Generally, systems which need resizing will be those which need a vacuum relief valve.
6. *Determine Minimum Flow Rate to Form Siphon*
   - a) This flow is that required to maintain at least 1 psi per 100 LF of friction drop in all the system high points working together. To be safe, for systems with several high points, use 1.5 or 2.0 psi per 100 LF.
7. *Size Supply Piping*
   - a) Use normal rules, that is, approximately constant pressure drop, and a friction loss rate of 2 to 4 psi per 100 LF.

8. *Calculate Supply Side Friction Loss*
   a) Calculate the loss—the static head—first for the minimum flow rate to form a siphon, found in Step 6.
   b) Then calculate the loss for the design flow rate.
9. *Calculate the Return Side Friction Loss*
   a) Calculate only for the design flow rate.
10. *Establish Fill-up System Curve*
    a) System curve at fill-up starts with the gravity head at zero flow rate, and runs through the two points of the supply-side curve found in Step 8.
    b) The intersection of this curve with the pump curve must occur at a flow rate equal to or greater than that found in Step 6, or the siphon will not form.
11. *Establish Midwinter System Curve*
    a) Find operating curve for system running at its coolest normal condition with siphon formed. Operating curves for hot conditions will lie between this curve and fill-up curve.
    b) Using the following formula, assume that the friction loss rate in the return line is constant in the vicinity of the end of the siphon (that is, there is no major constriction such as a group of elbows near the place where the siphon breaks. If such a condition exists, go back and refine the figures after the first estimate).
    c) ATH = atmospheric head at altitude of system, and at temperature assumed
    d) GH = gravity head from top of tank water to top of system
    e) FLR = friction loss rate in return piping, measured in feet per *foot*. For instance, for 4.32 psi per 100 LF, the FLR would be 10 ft/100 ft or 0.1.
    f) System head = supply side friction loss + GH − ATH/(1 − FLR)
    g) Calculate for another flow rate to establish the curve.
12. *Establish NPSA* (Net Positive Suction Head Available)
    a) Determine atmospheric head at highest *store* temperature allowed, and at altitude of project. (This is not the same atmospheric head used in step 11).
    b) Add to this the gravity head between the pump and the water level of the tank.
    c) Subtract from this the friction drop in the piping between the tank and the pump at design flow rate.
    d) The result equals the NPSA.
    e) If fluid other than water is used, consult other references.

13. *Select Pump*
    a) Pump must have no brass, aluminum, cast iron, or steel in the water stream. Use plastic, stainless steel, or bronze.
    b) Use identical pumps if more than one are used in tandem.
    c) If two tandem pumps are used, either double the flow rates on the pump curve, or cut system curve flow rates in half. For 3 pumps, use triple or one third; etc.
    d) Be sure pump selected has an NPSH at zero pressure drop less than the NPSA calculated in Step 12.
    e) Pump must be able to deliver the flow rate calculated in Step 6 (minimum flow to form a siphon) at the operating point where fill-up system curve intersects pump curve.
    f) Find operating point for system curve found in Step 11. This point, and the portion of the curve nearby and toward the fill-up operating point, should be at the optimum point on the pump curve. It should be near or slightly below the horsepower curve, not way above it (some manufacturers insist that the entire pump curve lie below the horsepower curve, for added safety factor). It should be near the highest efficiency of the pump. The pump curve and the system curve should intersect at close to a right angle, not at a very acute angle. Operating point should not be far out on the pump curve (at high flow rates), because if the system losses are overestimated the point may move toward even higher flow rates and might result in pump-motor overload or require a flow-throttling balancing cock in the discharge line, wasting pump energy.
    g) Operating point should be near design flow rate. This point is last on the list, because it is so difficult to find a pump which meets the other requirements, that meeting the exact design flow rate is usually impossible. Also, system losses sometimes vary over time, and vary with temperature, making an exact match unimportant. However, if possible, exceed the design flow rate rather than fall short, since the flow rate will go down as the system temperature rises (since the down-leg of the siphon will shorten).
    h) A warning: some manufacturers of specialty pumps do not publish accurate pump curves, and many do not publish NPSH data. Pumps with high speeds and small impeller "eyes" will have high NSPH requirements. *Sacrifice efficiency for low NPSH if necessary*. Pump efficiency is important, but is not the crucial design goal. The system will have a high C.O.P. in any case, and at least 40% of the energy used by the pump will end up in the water stream.

GUIDELINES FOR DESIGNERS: A SYSTEM DESIGN CHECKLIST (FOR ACTIVE SYSTEMS WITH OR WITHOUT PASSIVE SOLAR GAIN)

1. Optimize energy conservation first.

2. Design building to control solar gain, and divert or take advantage of prevailing winds.

3. Calculate all the loads which might be served
   - Water heating
   - Ventilation-air preheating
   - Swimming-pool heating
   - Space heating (after passive gain and occupancy heat are subtracted from gross heat load).

4. Arrange the loads in order, starting with the coolest load. What you want to know is how much load there will be typically available to serve at each successively higher store temperature.

5. Referring to Guidelines from Chapter 2, find out how much collector area is available, and calculate roughly how much energy might be available.

6. Compare this with the available loads. Take into account the difficulty in locating storage, in reaching a place to locate a load heat exchanger, and the size of the load (if too small, not worth serving). In general, pick off the lowest temperature loads and those easiest to serve before trying to serve difficult loads, such as space heating.

7. Based upon the available area and budget, select a size for the collector, and decide which loads are to be served.

8. Estimate the average operating ΔT between the store and ambient temperatures. This can be done by trial and error. First assume a store temperature; then calculate the net gain for a typical January day using that store temperature. See if at that temperature, the available energy can be distributed to the available loads within 18 to 24 hours. If not, adjust the store temperature until the supply and demand balance.

9. Based upon the expected operating ΔT, choose a working fluid, and select a collector type. Both the working fluid and collector design are very sensitive to operating temperature. If less than 75°F or so, simple relatively high-loss collectors can be used; if 100°F or more, relatively sophisticated collectors are necessary. With the collector type often goes the working fluid; for instance, many high-output collectors will not work with drain-back systems, because the collectors have been designed for "high-tech" installations with antifreeze.

10. Choose an actual collector among the various designs available in the chosen type. Using the expected ΔT, the average ambient for Janu-

ary, and the expected insolation on a clear day at about 10:30 A.M., find the average operating point for the collector in January. Based on the output and the actual price of the installed collector, compare cost effectiveness. Remember to adjust the real price for the collector for such features as internal manifolding, forming a roof, and requirement for heat-dump circuits, all of which will make one collector cost more or less than it seems relative to another. Be sure to use the price to a contractor in the quantity needed.

11. Select a storage tank or rock bin.
12. Lay out the basic piping runs, pump locations, and piping to the various load heat exchangers.
13. Lay out the array piping.
14. Size the load heat exchangers.
15. Size piping and select pumps.
16. Decide on control logic and select appropriate control devices.
17. Add all necessary details, and design the collector supports.
18. Add instrumentation. Try to sneak in a Btu meter on the load side whenever you can. It will tell you all you want to know, namely how much energy you are delivering to the load from the solar system. Also, put in plenty of thermometers and pressure gauges.

BIBLIOGRAPHY

AIA Research Corp. *Regional Guidelines for Building Passive Energy Conserving Houses.* Washington, D.C.: Superintendent of Documents, Document #HUD-PDR-355, Nov. 1978.

Despite its clumsy title, an idea-filled book with excellent illustrations. Using contemporary and historical examples, shows what to do in each of 13 climates, which are well-defined. A bit unselective and almost devoid of numbers.

AIA Research Corp. *Solar Dwelling Design Concepts.* Washington, D.C.: U.S. Government Printing Office, 1976.

A useful book for basic exposure to the solar design process. The author contributed the graphics and the designs in Chapter 5; the presentation of the sky dome and of solar interference contours are more detailed than in the present volume.

AIA Research Corp. *Survey of Passive Solar Buildings.* U.S. Department of Housing and Urban Development, Document #PDR 287, February 1978.

American Planning Association. *Site Planning for Solar Access: A Guidebook for Residential Developers and Site Planners.* Earley, D. and Jaffee, M. Washington, D.C.: U.S. Government Printing Office, Document #HUD-PDR-481, Sept. 1979.

The results of a HUD contract, this volume covers the subject of solar site planning in depth. Its illustrations are often inaccurate and amateurish. It focuses upon the ground plane shadow as a planning tool.

American Society of Heating, Refrigerating and Air Conditioning Engineers, Inc., (ASHRAE), *ASHRAE GUIDE.* In four volumes: *Systems* (1976); *Fundamentals* (1977); *Applications* (1978); and *Equipment* (1979); revised in rotation every four years. New York, ASHRAE.

Indispensible tools for solar heating system designers, particularly the *Fundamentals* volume, chapters 8, 19, 20, 21, 22, 23, 24, 26, 30, 31 and 32; and the excellent overview which comprises Chapter 58 of the *Applications* volume.

American Society of Heating, Refrigerating and Air-Conditioning Engineers, Inc. (ASHRAE). *ASHRAE Journal.* New York: ASHRAE, monthly.

Contains a continuing flow of solar articles.

Anderson, Bruce. *Solar Energy: Fundamentals in Building Design.* New York: McGraw-Hill Book Co. 1977.

A wide-ranging, general text with many references cited in the text. Reflects (as does the present book) the author's experience and prejudices, with an emphasis on air systems, high mass structures and passive. A perfect complement to the present book.

Baer, Steve. *Sunspots.* 2nd edition. Albuquerque: Zomeworks Corp. 1977.

An absolutely wonderful book exposing an inventive mind playing with physics, rather as Bucky Fuller did before him.

Clark, E. F. and deWinter, F., eds. *Proceedings of the First Workshop on The Control of Solar Energy Systems for Heating and Cooling.* Newark, Delaware: American Section of I.S.E.S. May 1978.

While filled with papers which ignore the forest for the trees, the papers on proportional control (pp. 80–89) are worth reading. Reading the other papers will help explain how simple truths have been lost in the shuffle.

Duffie, J. A. and Beckman, W. A. *Solar Energy Thermal Processes.* New York: John Wiley & Sons, 1974.

The basic text, presumably to be revised and superseded by Duffie and Beckman's newest book (to be published 1980). Many of the basic concepts in the industry are derived from this book, and one will find uncredited borrowings from Duffie and Beckman everywhere in the literature. Very technical; definitive on thermal processes within active collectors.

Forrester, Jay W. *Principles of Systems.* Cambridge, MA: Wright-Allen Press, Inc. 1968.

A "preliminary" edition of a proposed book, but one containing a step-by-step layman's decription of exponentials, feedback, and system control.

Franta, G. ed. *Proceedings of the 4th National Passive Solar Conference.* Newark, DE: American Section of the I.S.E.S. October 1979.

The 4th conference proceedings reflect growing experience and field data. This publication is nicely organized, and is compact. The author has not had time to review the papers.

Howell, Yvonne and Bereny, Justin A. *Engineer's Guide to Solar Energy.* San Mateo, CA: Solar Energy Information Service. 1979.

A useful book on practical applications of solar energy.

Jordan, R. C. and Liu, Benjamin, Y. H., ed. *Applications of Solar Energy for Heating and Cooling of Buildings.* New York: ASHRAE Publication GRP 170, 1977.

A basic source for insolation data and a good presentation of Liu and Jordan's approximations in Chapter V. Chapter II, by John Yellott, discusses radiation measurement. Chapter VIII, by Austin Whittier, is a highly technical discussion of active collector thermal and optical performance.

Koenigsberger, Ingersoll, Mayhew, Szokolay. *Manual of Tropical Housing and Building, Volume 1: Climate.* London: Longman Group, Ltd. 1977.

A comprehensive look at building design in hot climates. Look here first before re-inventing various thermal control systems.

Kreith, F. and Kreider, J. F. *Principles of Solar Engineering.* Washington: Hemisphere Publishing Corp., 1978.

A useful and basic text. It is highly technical, but since it covers so much ground, it is not exhaustive. Areas covered well are heat transfer, optics, concentrators, and process heating.

Mazria, Edward. *The Passive Solar Energy Book.* Emmaus, PA: Rodale Press; 1979.

A nice book covering passive design fundamentals. Perhaps it is appropriate to take the numbers in the tables at the end with a grain of salt. In Section V is a beautifully illustrated presentation of the sky-dome and the sky map ideas (see *Solar Dwelling Design Concepts*, where the author presented a modest version of these ideas in 1975).

Miller, H., Riordan, M., and Richards, D., eds. *Proceedings of the 3rd National Passive Solar Conference.* Newark, Delaware: American Section of the I.S.E.S. Jan. 1979.

A ponderous compendium, in which some papers stand out:

pp. 9–14: Peter Calthorpe. "The Solar Shadow: A Discussion of Issues Eclipsed." Deals very well with the interrelated issues of energy and density.

pp. 241–245: W. Scott Morris. "Natural Convection Systems with Storage." Describes large thermosyphoning air systems.

pp. 260–265: L. G. Berglund and A. P. Gagge. "Thermal Comfort and Radiant Heat."

pp. 314–317: Winslow H. Fuller and Daniel C. Lewis. "Overheating in Light and Heavy Mass Houses."

pp. 318–322: Larry Palmiter and Blair Hamilton. "A Comparison of Performance Factors for Passive Solar Heating." This paper raises the issue of marginal cost effectiveness by pointing out that no agreed-upon baseline is used to calculate passive performance.

pp. 395–440: Technical Session 4A — SIMPLIFIED DESIGN METHODS. Various papers containing worthwhile descriptions of several simple design tools.

pp. 713–719: R. W. Besant, R. S. Dumont, and G. J. Schoenau. "The Passive Performance of the Saskatchewan Conservation House."

Mitchell, J. W., Freeman, T. L. and Beckman, W. A. "Heat Pumps: Do They Make Economic and Performance Sense with Solar?" *Solar Age* July, 1978, pp. 24–28.

A clear analysis which pokes holes in most schemes for solar assisted heat pumps.

Olgyay, Victor. *Design with Climate*. Princeton, NJ: Princeton University Press. 1963.

A classic, must-read book for anyone interested in building design, solar or not.

Prowler, D., ed. *Passive Solar, State of the Art: Proceedings of the 2nd National Passive Conference*, 3 volumes. Newark, Delaware: American Section of the I.S.E.S. 1979.

Vol. 1: General description of many interesting buildings.

pp. 30–34: Don Watson. "Beyond Passive Design: Toward an Integrative Conservation Technology."

Vol. 2: Some interesting technical data, including some good papers on nocturnal cooling.

pp. 493–500: Sean Wellesley-Miller. "Weather Responsive Building Skins: Concepts and Configurations." Description of various MIT approaches to sun-control by new skin materials.

pp. 565–569: Gordon L. Moore. "Fitting a Handle and Scale on Thermal Time Lag." This investigator hit upon the analogy between an electric time constant and the thermal decay within a building at about the same time the author and his associates did. The present book was written before the author stumbled onto this lucid and well-written paper.

Vol. 3: Generally forgettable, but with some notable papers and poster reprints.

pp. 769–772: Wayne Place. "The Solar Resource: Motion and Amplitude." An interesting complement to the author's discussion of solar angles in Chapter Three.

pp. 840–849: Ralph Knowles. "Solar Ethics—Urban Form." A superb paper on the formal and ethical issues of solar envelopes.

Romer, R. H. *Energy: An Introduction to Physics*. San Francisco: W. H. Freeman & Co. 1976.

A superb text focusing on energy, and presenting complex ideas without complex math. It also contains excellent charts (mostly from 1973 or before) of energy consumption patterns.

Stromberg, R. P. and Woodall, S. O. *Passive Solar Buildings: A Compilation of Data and Results*. Sandia Laboratories Document SAND 77-1204, 1977. (available through N.T.I.S.)

Data on some early passive houses plus the classic paper by J. Balcomb, J. Hedstrom, and R. McFarland, "Passive Solar Heating of Buildings," originally given at the 1976 Albuquerque passive conference; and some rules of thumb proposed by J. Balcomb.

Tully, Gordon F. *Sun-Pulse II*. New York: McGraw-Hill Book Co., 1979. (with Energy Learning Systems).

A solar heating design simulation for use on a TI-59 hand-held calculator, consisting of a manual and three magnetic cards. The manual contains "sun-pulse" weather data for 26 climates.

GLOSSARY

absorber That portion of a solar collector in which solar energy is converted into heat.

absorptance The ratio between the energy absorbed by a surface to the energy impinging on the surface.

absorptivity The property of absorbing solar radiation and converting it to heat.

active system A solar heating or cooling system in which energy is transported from one component to another by non-solar means, such as electricity. (But what if the electricity is generated by solar?) The definition preferred by the author is a system involving separate, controllable elements which can be thermally isolated from the thermal load.

air mass The ratio of the distance through the atmosphere traversed by a sun ray to the distance through the atmosphere straight overhead at sea level.

altitude, solar The angle of the sun above the horizon.

ambient In general, the surroundings; ambient temperature is outdoor temperature when referring to buildings.

aperture In a collector, the opening through which the solar energy enters the collector.

ASHRAE The American Society of Heating, Refrigerating and Air-Conditioning Engineers, 345 E. 47th Street, New York, NY 10017.

auxiliary heating system The back-up heating system.

azimuth The number of degrees between true north and a vertical circle passing through the sun.

back-up heating system The normal fossil-fueled or electrical heating device which supplements a solar heating system.

black-body An ideal body or surface that completely absorbs all radiant energy falling upon it with no reflection.

British thermal unit Approximately, the amount of heat necessary to raise one pound of water from 59° to 60°F. The Btu is actually defined in reference to SI units.

elevation, solar The solar altitude.

emittance The ratio of the total radiant flux emitted by a body to that emitted by an ideal black body at the same temperature.

emissivity The capacity of a material to emit radiant energy.

energy A term which is virtually impossible to define precisely. Energy in general is one of a number of properties of matter which is conserved. Thermal energy is of particular importance in solar heating, since it defines the kind of energy resulting from rapid, random motions of atoms and molecules of the substance, an increase of which raises its temperature. Solar energy is a form of electro-magnetic energy which is transmitted from place to place by means of photons. For a fuller discussion of these issues, see Romer's "Energy: An Introduction to Physics." (See Bibliography.)

equinox Either of the two times each year when a line from the center of the sun to the center of the earth crosses the equator, and day and night are everywhere of equal length. Equinoxes occur about March 21st and September 23rd.

greenhouse effect The suppression of heat loss by the re-reflection of radiation, as occurs in the atmosphere. It is inaccurate to apply this term to flat plate collectors, except as a secondary effect of placing a cover over the absorber. The primary effect of the cover is to prevent convective heat loss

gross area The overall area of a definable solar collector. Using gross area to measure performance allows one to compare a rectangular flat plate collector with discontinuous devices such as linear tracking collectors, or vacuum tubes which are spaced apart. The gross area normally does not include manifold piping.

heat capacity rate For a moving fluid, the volumetric flow rate times the density times the specific heat.

heat exchanger A device to transfer heat from one fluid to another while keeping the fluids separated.

heat loss coefficient The heat transmission coefficient (see U-factor).

heat of fusion The latent heat involved in a change of state from solid to liquid.

heat of occupancy The heat generated within a space incidental to the occupancy of the space. Also called *miscellaneous heat, wild heat, free heat,* etc.

heat pump A refrigerating system employed to transfer heat from one space or substance to another. Heat is picked up by the evaporator, transferred through the working fluid and discharged through the condenser.

hydronic Related to heating and cooling with liquids.

incident angle The angle between a ray of light striking a surface and a perpendicular to that surface.

indirect gain system A passive solar heating system in which solar energy passes through a glazing cover and falls directly on a storage device, without entering the space to be heated.

infiltration The unintentional flow of air into a space through cracks, open doors, etc.

infrared radiation That part of the electromagnetic spectrum which exceeds 0.7 microns; heat.

insolation Incident solar radiation; the irradiance from the solar source.

intrinsic storage As used by the author, the thermal storage capability of the building materials themselves, such as the thermal slab, gypsum wallboard, interior masonry, etc.

latent heat Technically, the change in enthalpy during a change of state (enthalpy is roughly equivalent to the obsolete term heat content). As used in solar applications, it is the heat absorbed or emitted by a substance without changing temperature when undergoing a change of phase from a liquid to a solid or a liquid to a gas.

load factor The ratio of the actual mean load to a maximum load or maximum production capacity in a given period.

magnetic declination The angle between compass north and true north for any given point.

MBtu In some contexts, 10^6 Btu (mega-Btu); in others, 10^3 Btu (from the Latin $M = 10^3$). To be clear, use MMBtu to mean 10^6.

miscellaneous heat The heat of occupancy.

net area The aperture, when referring to solar collectors.

net positive suction head The pressure required on the suction side of a pump at the eye of the impeller to prevent the fluid in the pump from vaporizing, at any given temperature. NPSH refers to the requirement; NPSA refers to the head available in the system.

normal In the context of sun angles, perpendicular.

off-angle correction An adjustment made to the value of the incident radiation on a glazed surface to account for the increased reflection off the surface at high incident angles.

orientation The azimuth of a line normal to the surface of a solar collector, measured from true south.

passive Commonly, any solar heating process which uses personal or solar energy to transport heat from one place to another. In a more general sense, describes solar applications which involve the enhanced use of the sun's energy in buildings by the manipulation of materials and configurations, other than by means of added, thermally separable elements.

payback period (simple) The inverse of the first year return on investment. A crude but satisfactory check on solar system feasibility.

penetration level The fraction of a given load provided by any particular form of energy generation. For solar energy, the solar fraction.

phase change A change from one of the fundamental states of matter (solid, liquid, gas) to another, always involving the absorption or emission of latent heat.

photovoltaic cell A semi-conducting device which produces a direct current of electricity in response to radiant energy incident upon its surface.

power The rate at which energy is transferred.

profile angle The angle formed with the horizontal by a projection of the sun's rays on a plane perpendicular to the surface of the collector. For example, in a collector facing true south, the profile angle is the apparent angle between the sun's rays and the ground when the observer is facing true east. At noon, the profile angle equals the sun's elevation.

pyranometer An instrument for measuring the total incident solar radiation.

quad One quadrillion Btu's (10^{15}). Do not confuse with another scientific unit Q, which is 10^{18} Btu's. One quad equals one mQ. Moral: do not abbreviate quad with Q! U.S. energy consumption is in the vicinity of 75 quads per year.

R-factor The inverse of the U-factor. The R-factor measures thermal resistance as opposed to thermal conductance. R-factor's can be added; U-factors cannot be added.

reflectance The ratio between the radiant energy reflected from a surface and the incident radiant energy.

reflectivity The property of reflecting radiant energy.

retrofit The installation of solar or energy conservation equipment (for instance) on existing structures.

selective surface A surface which exhibits a high absorptivity in the visual range of the spectrum, but a low emissivity in the infrared. Used as a coating for absorber plates in solar collectors.

sky dome A term to describe an imaginary hemisphere centered on the observer upon which can be drawn the sun's paths through the sky at various times of the year.

solar constant The solar intensity incident on a normal surface located outside the earth's atmosphere at a distance from the sun equal to the mean distance between the earth and the sun. Equal to 1353 W/m^2, or about 429 $Btu/hr \times ft^2$.

solar fraction That portion of a given load supplied by solar energy.

solar rights The legal right of a property owner to prevent the obstruction of useful sunlight falling upon his property.

solar thermal Relative to low temperature systems, any process in which solar radiation is converted into thermal energy. The term is more commonly used in regard to high temperature systems, in which radiant energy is first converted to high temperature thermal energy, for use in generating electricity or in driving industrial processes. To understand which meaning is intended, it is important to watch the context carefully.

solstice Those two points in the year, occurring about June 22nd and December 22nd, at which the sun's apparent north to south motion in the heavens comes to a stop and reverses. The longest and shortest days and nights occur on the solstices.

specific heat The ratio of the quantity of heat required to raise the temperature of a given mass of any substance one degree to the quantity required to raise the temperature of an equal mass of a standard substance (usually water at 59°F) one degree. Water, therefore, has by definition a specific heat of 1.0.

stagnation That condition reached frequently in an active solar collector in which all of the incident solar energy is dissipated to the environment. The term is commonly used to refer to stagnation during periods of high insolation and high ambient temperatures, when the solar collector would typically be under severe thermal stress.

stratification The condition of a storage device in which thermal energy is "layered," forming a temperature gradient.

thermal conductivity The time rate of heat flow through a unit area and unit thickness of a homogeneous material under steady conditions when a unit temperature gradient is maintained in the direction perpendicular to the area.

thermostat An instrument which responds to changes in temperature and which directly or indirectly controls the temperature in a feedback loop.

thermosyphoning The natural convection of a fluid under the influence of a temperature gradient and the force of gravity.

tilt The angle formed by the plane of a solar collector and the ground.

ton of refrigeration The removal of heat at the rate of 12,000 Btu's per hour (2,000 lb × 144 Btu/lb × 24 hours, where 144 Btu/lb is the latent heat of fusion of ice).

transmittance The ratio of the radiant energy transmitted through a body to that incident on it.

transmissivity The capacity of a material to transmit radiant energy.

turbidity A measure of the particles and gasses in an atmosphere which cause it to become obscure.

U-factor The rate at which heat flows through a unit area of a barrier, under steady conditions, from the fluid on the warm side to the fluid on the cool side, per unit temperature difference between the two fluids. Also called the heat loss coefficient or thermal transmittance.

ultraviolet That region of the electromagnetic spectrum which has shorter wavelengths than visible light (less than 0.4 microns). While technically covering the entire region of short wavelength radiation, in solar energy it is usually confined to the near UV region, in which occurs part of the sun's radiant spectrum.

work Technically, the component of a force in the direction of motion times the distance traveled by the object on which the force acts. Work, like energy, is an illusive concept to define; but at the heart of the concept of work is its *usefulness*. The second law of thermodynamics describes important relationships between useful work and thermal energy. Heating systems do no work, but the pumps and fans perform work.

TABLE OF CONVERSIONS

English to SI

Length

*1 in = 2.54 cm

*ft = 0.3048 m

Area

*1 ft^2 = 0.093 m^2

Volume

1 ft^3 = 28.32 liter

1 U.S. gallon = 3.785 liter

1 cfm = 0.4719 liter/s

1 U.S. gpm = 0.0631 liter/s

Mass

*1 lb = 0.45359237 kg

1 lb/ft^3 = 16.02 kg/m^3

Energy

1 Btu = 1.055 kJ

*1Btu/lb × °F = 4.1868 kJ/kg × °C

*1 Btu/lb × °F = 1 kcal/kg × °C

*1 Btu/lb × °F = 1.163 Wh/kg × °C

Power

1 Btu/hr = 0.2931 W

10^6 Btu = 293.1 kWh

1 Btu/hr × ft^2 = 3.155 W/m^2

1 Btu/hr × ft^2 × °F
= 5.678 W/m^2 × °C (U-factor)

1 ton refrigeration = 12,000 Btu/hr
= 3.517 kW

1 hr × ft^2 × °F/Btu
= 0.1761 m^2 × °C/W (R-factor)

Temperature

1°F ≡ 0.556°C

*1°F ≡ 1°R

0°F = 459.7°R

0°F = −17.78°C

Thermal Conductivity

1 Btu/hr × ft^2 × (°F/in)
= 0.1443 W/m × °C

SI to English

Length

1 cm = 0.3937 in

1 m = 3.281 ft

Area

1 m^2 = 10.764 ft^2

Volume

1 liter = 0.0353 ft^3

1 liter = 0.2642 U.S. gallon

1 liter/s = 2.119 cfm

1 liter/s = 15.85 U.S. gpm

Mass

1 kg = 2.205 lb

1 kg/m^3 = 0.06243 lb/ft^3

Energy

*1 W = 1 J/s

*1 cal = 4.1868 J

*1 kWh = 3.6 MJ

*1 Wh = 3.6 kJ

1 kJ/kg × °C = 0.2388 Btu/lb × °F

1 Wh/kg × °C = 0.8598 Btu/lb × °F

Power

1 W = 1 J/s

1 W = 3.412 Btu/hr

1 kWh = 3.412 kBtu

1 W/m^2 × °C
= 0.1761 Btu/hr × ft^2 × °F

1 m^2 × °C/W
= 5.678 hr × ft^2 × °F/Btu

Temperature

*1°C ≡ 1.8°F

*1°C ≡ 1°K

0°C = 32°F

0°C = 273.15°K

Thermal Conductivity

1 W/m × °C
= 6.931 Btu/hr × ft^2 × (°F/in)

*exact conversion

INDEX

Absorption
  losses 45
  molecular 29
Access, solar 14, 22, 24
Air mass 28
Angle
  hour 36
  of incidence 14
  profile 17
  sun 9
Appliances, heating effect of 112, 164
Attic, solar 21
Auxiliary (*see* Back-up)
Averaging of sun-pulse amplitude 72

Back-up system
  control of 142
  hydronic 143
  optimizing 143, 181
  warm-air 142
Base load 122, 142
Break-even condition 67, 75

Capacitance 99
  of storage 123
Capacitance rate (Gc_p) 75
Carry-over 8, 133
Cavitation, pump 197
Clouds 31
  "popcorn" 34, 77
Collectors, active
  absorption losses 45
  air 61, 148
  black fluid 54
  concentrating 54
  construction of 51
  control of 67, 73, 148
  cycling 74, 79
  effect of wind on 63
  efficiency 48
  efficiency factor (F_R) 59
  fin efficiency 57
  glazing 45, 50, 66
  heat loss 47
  heat loss coefficient ($F_R U_L$) 62
  heat transfer coefficient 54
  heat removal factor 60
  insulation 50
  kt product 57
  low-loss 32, 163, 193
  massive 71
  mounting 21, 50
  operating temperature 3, 71
  output equation 44, 62

Collectors, active *(Cont.)*
performance curves 62, 69
selective surface 52
sizing 180
stagnation temperature 63, 74, 193
tests 48, 69
thermal shock in 193
tilt 17, 39, 125, 129
transient response to weather 32, 71, 77
transmission losses through cover 45
tube and sheet 57
vacuum tube 32, 52
Conservation, energy
and occupant behavior 110
and water heating 163
competition with solar 132
before adding solar 175
Constant, solar 27
Consumption, energy (*see* Load)
Control (*see also* Thermostat)
cycling 74
high-limit 135
low-limit 137, 147
of air collector 148
of back-up system 142
of collector 67, 73
of drain-back system 194
of heat pumps 158
of supply 122
proportional 79
Cost-effectiveness (*see* Payback)
and solar access 16
determining cost 175
determining output 176
in heat exchanger selection 95
in passive systems 48, 154, 178
marginal (incremental) 178
measure of 173
system 127
Curves
collector performance 62, 69
horsepower 203
pump 197
system 202
Cycling in collector operation 74, 79
Declination, compass 12, 20
Degree-day 117, 131
Demand, heating (*see* Load)
Density, population 15
Deviation (*see* Declination)
Differential controller 173
Drain-back system
components of 146, 157
control of 194
draining of 191
freeze-up of 193
operation of 188
oxygen in 192, 196
piping of 200
principles of 186
pumps 196, 200
storage tank design 198
valves in 194
venting 186, 198
water heating 169
Drain-down systems 169, 192

Effectiveness 85, 157
Efficiency
collector 48
factor, collector 59
fin 57
instantaneous water heater 162
tests 48, 67
Electric consumption (*see* Load)
Equinox 10
Exponential decay 97

Film
air 105, 109
coefficient 85, 109
Flow rate
hot water 167
in collector 59, 180
in system design 201
F_R (*see* Heat removal factor)
Fraction, solar 16, 181 (*see* Penetration Level and Supply and Demand)

Gain, solar
east and west 23
in passive designs 151, 158
summertime 23
through exterior wall 108
Glazing
collectors 45, 50, 66
passive 150
window 109
Greenhouses 151

Head
atmospheric 188
gravity 189
pump shut-off 198
static 191
system 188
Heat capacity rate ($\dot{m}c_p$) 84, 91
Heat dumping 153, 179
Heat exchanger
baseboard 139
building considered as 83, 104
collector considered as 47, 54
definition 83
rock store considered as 147, 157
types 86
Heat loss
calculating 117
collector 47, 62, 73
from storage 136
rate 106, 157
system 3
through glass 109
through walls 105
to ground 109, 114
Heat pump 158
water heater 164
Heat removal factor 60, 147, 180, 201
Heat transfer coefficient, collector 54
Humidification 113
Hundred percent (100%) solar
periods 132, 136
systems 130, 133

Immersion coil 1, 90, 147
Inertia, thermal 34, 122, 140
Infiltration 84, 109
Insolation (*see* Sun-pulse and Supply and Demand)
and collector sizing 204
definition 45
diffuse 14, 28
direct beam 13, 28
hourly 26
peak versus daily 33
tilted, calculating 29
Instrumentation 177, 205
Insulation
collector 50
definition 105
of slab edges 115
optimizing 178
piping, and controls 79
storage tank 199
water heater 165
Interference, solar (*see* Access, solar)
$\overline{K}_T$, K_T, k_T 29
kt product 57

Lighting 158
heating effect of 112, 116
Load
base 122
clothes drying 119
dead 21
diversified electric 118
gross 126
hot water 127, 167, 181
management 123
net 126
relation to supply 121
space heating 126
washing 119
wind 21

Mass
air 28
and collector response 71
effect on nighttime set back 140
in collector control 77
in passive design 153
Mass wall 151, 155

North
  plan 21
  Star 12
  true 12
NPSH and NPSA 189, 197, 202

Occupant behavior 109
  and water heating 164
Optimizing (*see* Cost-effectiveness and Payback)
  collector size 180
  effect of low limit on 139
  effect of nighttime setback on 141
  flow rate in collector 59, 179
  heat exchanger size 95
  hot water tank size 165
  hydronic back-up 143
  insulation thickness 178
  passive storage 178
  spacing between collectors 17
  store and heat exchanger together 101, 179, 182
  storage tank size 179
  systems 172
Orientation, collector 17, 20, 37
Oxygen, in drain-back system 192, 196

Parasitic losses 179
Passive systems
  and collector heat loss 47
  and output of vertical collectors 130
  and sun-pulses 35
  determining cost of 175
  direct gain 151
  optimizing storage of 178
  types 150
Payback 48, 174 (*see* Cost-effectiveness and Optimizing)
Penetration level, solar 14, 17 (*see* Supply and Demand)
Phase-lag storage (*see* Storage, seasonal)
Piping
  buried 21
  design of, in drain-back system 200
  pitching of 21, 193
  return, size of 189
  temperature 193
Pump
  curves 197
  in drain-back system 188
  motor, start-up in reverse 194
  operating point 203
  parasitic losses 179
PV cells 23
Pyranometer 25

R-Factor 106, 178
Radiation (*see* Insolation and Sun-pulse)
  and collector heat loss 64
  black-body 27
  extraterrestrial 27
  from fireplace 107
  in heat transfer 105
  in orbit 31
  infrared 28
  ultraviolet 28
  visible 28
Reflection
  ground 39
  off angle 25, 35
Reflectors 41
Refractive index 37
Refrigerator (*see* Appliances)
Reservation, solar space (*see* Access, solar)
Rock stores 146, 157

Scattering, molecular 29
Selective surface 52
Shading
  checklists 21, 23
  effect on controls 79
  on collector 14
  self- 18, 24
  sun-pulses 37
  trees 22
  with off-south collectors 37
Shock, thermal 193
Simulation
  and storage *vii*
  of occupant behavior 111

of weather data 42
time intervals 119
use in determining output 177
Siphon, in drain-back system 188
Site planning 18, 23
Sky dome 11, 29
Sky map 11
Snow, on collectors 23, 129
Solstice 10, 21
Spectrum, solar 28
Stack effect 109
Stagnation 63, 74, 193
Storage
as primary system variable *vii*
heat-of-fusion 155
high-temperature limit in 135
in direct gain passive systems 153
in water heating systems 162
rock 146
seasonal 134
skin losses 136
tank, hot water 165
tanks, buried 21
tanks, design 198
when needed 6
Stratification
in buildings 110
in rock stores 146
in water heaters 163
in water storage 7, 144
Sun-pulse
clear-day 25
cloudy-day 31
concept, summary 41
definition 6
derivation 27
non-south 37
relating to collector performance 70
sequence of 132
shaded 37
Supply and demand curves 126
for 100% solar periods 132
for seasonal storage 134
for water heating 161
Swimming pool heater 47

Tanks, storage (*see* Storage)

Temperature
ambient 31
and low-limit control 138
break-even 67, 75
collector, operating 3, 71
delivery 7
hot water use 165
relation to R-Factor 107
stagnation 63, 74, 193
surface diagram 57
system 3
system decay 96
turn-on differential 77
variation as function of supply 122
Thermostat
morning set-up 142
nighttime set-back 139
setting, relation to annual load 130
two stage 142
Tests, collector efficiency 48, 69
Thermosyphoning water heaters 169
Tilt, collector 17, 39, 125, 129
Time constant 98, 121
Time, solar 37
TMY data 29
Transition months 126, 133 (*see* 100% solar)
Transmission losses through glazing 45
Transient response of collector 32, 71, 77
Trees
barriers and windbreaks 24
deciduous, shading 22
Trombe wall 179 (*see* Mass wall)
Turbidity, atmospheric 29

U-Factor
collector 48
heat exchanger 83
in buildings 106
relation to time constant 100

Vacuum relief valve 192
Valves
in drain-back systems 192, 194

Ventilation 109, 111
  under roofs 50
Venting in drain-back system 186, 198

Water heating 161
  and consumption 167
  and energy conservation 163
  and heat pumps 163
  individual metering 165
  instantaneous 161
  off peak 163
  recirculating loops 165
  system types 169
Weather data 29, 42
Wind, effect on collector 63